RAND NATIONAL DEFENSE RESEARCH INSTITUTE

Cost-Effective Helicopter Options for Partner Nations

Christopher A. Mouton, David T. Orletsky, Michael Kennedy, Fred Timson, Adam Grissom, Akilah Wallace

Prepared for the Office of the Secretary of Defense
Approved for public release; distribution unlimited

For more information on this publication, visit www.rand.org/t/rr141z1

Library of Congress Control Number: 2015931890
ISBN: 9780833087164

Published by the RAND Corporation, Santa Monica, Calif.

Preface

The Department of Defense (DoD) provides significant assistance to partner nations in support of counterterrorism and counterinsurgency operations. The partner nation's helicopter fleet is a critical component of these operations. Many important security partners have helicopter fleets that are a legacy of the Cold War. Since DoD provides significant resources to support these aircraft, an assessment is required to determine if transitioning to different aircraft is more cost-effective than continuing to maintain an aging fleet. This document reports the findings of a cost-effectiveness analysis of helicopter platform options for partner nations.

This research was sponsored by the Office of the Secretary of Defense and conducted within the International Security and Defense Policy Center of the RAND National Defense Research Institute, a federally funded research and development center sponsored by the Office of the Secretary of Defense, the Joint Staff, the Unified Combatant Commands, the Navy, the Marine Corps, the defense agencies, and the defense Intelligence Community.

For more information on the International Security and Defense Policy Center, see http://www.rand.org/nsrd/ndri/centers/isdp.html or contact the Director (contact information is provided on web page).

Contents

Figures

Tables

Summary

Department of Defense (DoD) assistance to partner nations often entails supporting their helicopter fleets. In some cases, these fleets are composed of nonstandard rotary-wing (NSRW) aircraft, usually Soviet-era or Russian, European, Chinese, or outdated American equipment. Partner use of NSRW aircraft poses particular problems for U.S. security cooperation activities; understandably, the U.S. military does not have a large base of expertise to support these aircraft in such areas as flight crew training, maintenance, and supply chain management.

Recognizing these problems, DoD asked RAND's National Defense Research Institute to identify the aviation requirements of important partner nations,[1] apply this understanding to an analysis of the relative efficiencies of a variety of helicopter platforms, and use these findings to quantify the cost-effectiveness implications of transitioning partner nation NSRW fleets to alternative aircraft. In this report we present the method and results of the second and third of these tasks.

In this summary, we provide a brief synopsis of the analytical approach used to generate these results, then present our findings in greater detail.

The RAND Analysis

The first step of the RAND analysis was to set the scope of the cases to be included for analysis. We chose to focus particularly on the helicopter requirements of four key partner nations—including Afghanistan and Iraq—plus 29 others.

To make the analysis of so many partners tractable, we differentiated the total of 29 into common groups of countries based on likely mission distance and altitude. This produced a clustering around missions flown at low altitudes for short and medium distances—less than 100 nautical miles. We conducted our cost-effectiveness analysis for all nine combinations of distance and altitude. However, for simplicity, we will focus on the analytical boundaries of the mission space: high-altitude/short-distance; high-altitude/long-distance; low-altitude/long-distance.[2] For each of these altitude-range combinations, we defined (and used in our analysis)

[1] These results can be found in Adam Grissom, Alexander C. Hou, Brian Shannon, and Shivan Sarin, *An Estimate of Global Demand for Rotary-Wing Security Force Assistance*, Santa Monica, Calif.: RAND Corporation, 2010, not available to the general public.

[2] Detailed results for the other key partner and archetype nations are presented in a companion piece, not available to the general public: Christopher A. Mouton, David T. Orletsky, Michael Kennedy, Fred Timson, Adam Grissom, Akilah Wallace, *Cost-Effective Alternatives to the Mi-17 for Partner Nations: Focus on Afghanistan, Iraq, Pakistan, and Yemen*, Santa Monica, Calif.: RAND Corporation, 2014.

an archetype nation, such that the results of the cost-effectiveness modeling are applicable to all countries without being specific to any one. Each of these categories with its constituent number of partner nations is displayed in Table S.1.

The second step was to identify the helicopter alternatives to be considered. We analyzed 28 helicopters: the Mi-17v5, which we take as representative of current partner nation fleets, and 27 potential alternatives. The alternatives are a mix of utility (lift) and attack helicopters, military and civilian, and are of various sizes, as listed in Table S.2. We evaluate each alternative helicopter against the Mi-17v5 using a methodology that allows us to arrive at a meaningful measure of cost-effectiveness. This is achieved through a modeling process that integrates best-estimate costs—how much money must be spent per aircraft—with helicopter performance. Importantly, we quantify helicopter performance not as a product of mechanical specifications, but rather as the product of the ability to complete mission-tasks that are representative of those being undertaken, or likely to be undertaken, in U.S. partner nations.

The third step in the cost-effectiveness methodology is to identify the capabilities partner helicopters need if they are to achieve U.S. security objectives. The purpose of U.S. assistance to these partner nation helicopter units is to enable them to conduct operations that further the interests of both the partner nation and the United States. We therefore relied upon U.S. military and DoD personnel to establish the set of necessary and sufficient mission requirements. These are defined in terms of *mission-tasks*, a measure of what a helicopter fleet must be able to do. By connecting mission requirements to helicopter capability, mission-tasks link operational needs directly to fleet size for each helicopter analyzed.

Mission-tasks are not meant to be predictive, they are merely realistic representations of the type of performance that partner fleets can be expected to need in the future. We used a wide range of information (described in detail in Chapter Two) to help us define mission-tasks such that they determine the total amount of personnel and equipment to be moved, the time frame of the movement (e.g., single lift, one period of daylight), the departure base, and the landing zone, but they do not dictate the number of passengers on each sortie.[3] These data

Table S.1
Grouping of Partner Nations for Analysis

		Typical Mission Distances			
		Short (10nm – 50nm)	Medium (50nm – 100nm)	Long (100nm – 200nm)	Total
Typical Landing Zone (LZ) Altitudes	Low (1,000' – 4,000')	6	10	4	20
	Medium (1,000' – 10,000')	3	1	1	5
	High (6,000' – 18,000')	1	1	2	4
	Total	10	12	7	

[3] Defining the minimum number of passengers on a sortie should be avoided, unless there is an operationally significant reason to do so, because it can cause alternatives to fail and be eliminated from further consideration. For example, if the task is to deploy a company of 100 personnel in a single lift, is it critical to have 20 personnel per aircraft (resulting in five sorties) or would 15 passengers per aircraft be acceptable? It is clear there is some minimum number of passengers that is operationally significant, and minimums should be established to eliminate completely unreasonable cases. It is best, however, to set this limit low so that the most alternatives are carried through the analysis.

Table S.2
Alternative Helicopters Considered

Military Utility / Lift	Civilian Utility / Lift	Civilian / Military Utility / Lift	Attack
Huey II	Bell 407	S-92A	AH-1Z
UH-1N	Bell 429	AS-332L1	AH-46D
UH-1Y	Bell 412EP	S-61T	AW129
CH-46E	AW109	S-61T+[b]	Mi-35
CH-47D	EC-145[a]	AW139	
CH-47 Int		Mi-17v5	
AW149			
UH-60L			
UH-60M			
NH-90			
EC-725			
EH101			

[a] The LUH-72A Lakota is a variant of the EC-145.

[b] The S-61T+ is a new Sikorsky aircraft currently in the design phase that is designed to compete with the Mi-17v5. HIP-H is the older version of the Mi-17.

allowed us to calculate mission distance, and after accounting for en route terrain and altitude, produced a set of 100 routes. Mission-tasks also include a description of the temperature conditions under which the movement must take place, a variable that can influence greatly the performance of each helicopter.[4] Together, these parameters produce a set of rotary requirements that are country-specific and include a wide range of missions, including movement of forces, sustainment of forces, medical evacuation, and casualty evacuation.

With the mission-tasks thus defined, the fourth step in the cost-effectiveness methodology is to analyze each helicopter's effectiveness. To arrive at a meaningful measure of each alternative's effectiveness relative to the Mi-17v5, we analyzed multiple mission-tasks on each of the 100 routes, at each of three climate conditions, for a total of 338 effectiveness values per alternative helicopter, per temperature—that is, more than 1,000 effectiveness values for each helicopter.

---

[4] Since it is difficult to apply specific weights to each temperature condition, a simple average was used. For example, an extremely hot day may only occur 1 percent of the year, but giving it a 1 percent weight may not be accurate because there may be great value in being able to operate on any day and at any time. Specifically, having this ability would deny the enemy known windows of immunity from rotary-wing assets. Based on this, in the authors' judgment, neither a 1-percent weight nor a 100-percent weight for extra-hot is appropriate, and therefore, for simplicity, we used weights of 33 percent for each temperature condition.

The effectiveness values are the ratio between the number of Mi-17v5s and the number of the alternative aircraft required to carry out the given mission-task: the quantity of Mi-17v5s divided by the quantity of the alternative. Thus, if a task can be completed with the same number of the alternative aircraft as Mi-17v5s, the effectiveness value, or score, of the alternative is 1.0—the two are equivalent in military capability. But if the number of the alternative aircraft required is half the number of Mi-17v5s, the effectiveness score of the alternative is 2.0; i.e., it has twice the military capability of an Mi-7v5, and so forth (Table S.3).

Knowing the effectiveness scores enabled us to use U.S. government, contractor, trade press, and other open-source data in our fourth step, in conjunction with survey responses from experienced military personnel, to arrive at a best-estimate of the cost of each platform, again relative to the Mi-17v5. The total life-cycle cost of an aircraft depends on both the flyaway cost of the aircraft and its operations and support (O&S) cost.

We obtained flyaway cost data from various outlets, including the U.S. government (USG), contractors, and trade press and other open-source media. For the Mi-17v5 we relied upon recent contracts and databases maintained by the USG, as well as two USG reports.[5] We requested price information from all of the contractors producing the helicopter alternatives included in our analysis, and received responses from Bell, Boeing, Sikorsky, and Eurocopter. Where available, we accepted the valuation of the helicopters' commercial variants as provided by aviation consultancy Conklin & de Decker. In cases where even this information was unavailable, we used a "comparable data" approach; we were unable to arrive through any of these methods at a reliable estimate for three helicopters, the AW129, the Mi-35, and the Mi-17 (HIP-H). USG data were included in the model for all aircraft for which they were available, unless the contractor-provided estimate was higher.[6]

While source data on procurement price were generally consistent, the same cannot be said for O&S costs. There is great variation in how dollar-per-flying-hour O&S estimates are calculated. Although some data repositories—for example, Conklin & de Decker—do provide consistent estimates of O&S costs for a number of platforms, their accounting does not include the full set of helicopter alternatives being analyzed here. We therefore use the May 1, 1992, Office of the Secretary of Defense Cost Analysis Improvement Group (OSD-CAIG) structure

Table S.3
Effectiveness Criteria

Relative Capability	Effectiveness Score
10 alternatives have the mission capability of 20 Mi-17v5s	2.0
20 alternatives have the mission capability of 20 Mi-17v5s	1.0
40 alternatives have the mission capability of 20 Mi-17v5s	0.5

5 DoD, *Mi-17 Helicopters,* March 23, 2010b; Department of State (DoS), Aviation Services Study, March 9, 2009.

6 We saw no evidence to suggest contractor prices were biased, either up or down.

to define the categories of O&S to be included in our model.[7] We use the AFTOC structure as the basis for our model, supplementing with Army and Navy/Marine Corps helicopter maintenance data when available.[8] This structure is designed to capture the characteristics of the helicopters in use by the U.S. military; as such, we did not assume it would be equally applicable to the Mil Moscow Helicopters included in the RAND analysis, the Mi-17v5 and the Mi-35.

Just as with effectiveness, we operationalized cost scores as the ratio between the baseline Mi-17v5 and the alternative platform. This means that a *cost-score* is the unit cost of the Mi-17v5 relative to the unit cost of the alternative. If the cost of the alternative is the same as that of the Mi-17v5, its cost-value is defined as 1.0. If its cost is twice that of the Mi-17v5, its cost-value is defined as 2.0, and so on.

The fifth and final step of our analysis was to calculate each helicopter's cost-effectiveness compared to that of the Mi-17v5. Here, we define *cost-effectiveness* for each alternative as the ratio of its effectiveness score to its cost score. The resultant value is again in terms of the Mi-17v5—that is, if the alternative has half the cost and half the military effectiveness of the Mi-17v5, or if it has twice the cost and twice the military effectiveness of the Mi-17v5, then its cost-effectiveness is 1.0. Put another way, spending a given budget on Mi-17v5s or on the alternative gives the same level of military capability. If an alternative has the same capability as the Mi-17v5 and costs half as much, or if it has twice the capability of the Mi-17v5 and costs the same, its cost-effectiveness is 2.0—i.e., spending a given budget on the alternative gives twice the military capability as spending it on the Mi-17v5, and so forth. Table S.4 gives a sample accounting of this relationship; equivalence is indicated in green, and superiority in blue.

Key Findings

Several observations were made about the performance and cost-effectiveness of the reference helicopter, the Mi-17v5. It was unable to complete all of the mission-tasks on all the routes evaluated. In some cases, limitations were imposed by range, in others, by required altitude or

Table S.4
Relative Cost-Effectiveness

		Relative Effectiveness		
		0.5	1.0	2.0
Relative Cost	0.5	1.00	2.00	4.00
	1.0	0.50	1.00	2.00
	2.0	0.25	0.50	1.00

[7] This structure is consistent with that of the U.S. Air Force Total Ownership Cost (AFTOC). Office of the Secretary of Defense Cost Analysis Improvement Group, *Operating and Support Cost-Estimating Guide,* May 1, 1992.

[8] U.S. Army Operating and Support Management Information System (OSMIS) website, undated. U.S. Navy Visibility and Management of Operation and Support Cost (VAMOSC) website, undated. These websites are not available to the general public.

hover capability. We note also that the Mi-17v5 generally declined in cost-effectiveness relative to the alternatives as temperature increased.

Our analysis indicates that among utility platforms, the Boeing CH-47D, Sikorksy S-61T, Eurocopter AS-332L1 Super Puma, and the AgustaWestland AW139 are consistently more cost-effective than the Mi-17v5. The Sikorsky S-61T+ performs similarly well, but is in development. In these cases, the margin of increase in cost-effectiveness over the Mi-17v5 often is such that these aircraft achieve greater efficiency even when additional tail requirements are applied.

Several small utility helicopters also had good cost-effectiveness, including the Eurocopter EC-145 (LUH-72A Lakota) and the AgustaWestland AW109. Note, however, that these aircraft did not carry two loadmaster/gunner personnel in this analysis, nor a door-gun and ammunition. Thus, these platforms lack the defensive capability needed to suppress enemy action in the landing zone. The cost-effective results presented here do not penalize for this fact, so the findings must be interpreted with this in mind. For attack aircraft, the AH-1Z and the AW129 are able to accomplish many of the same routes as the Mi-17v5. That makes these aircraft feasible candidates for providing armed escort, and for overwatch of mobility helicopters.

Acknowledgments

The authors are grateful to many people for providing assistance on this research. We would first like to thank the DoD "tri-leads," Dr. John Crino, Lt Col Trevor Benitone, and Col Scott Frickenstein, who provided ongoing support and continuous assistance throughout our effort. We would also like to thank CDR Daniel Pheiff, who provided his operational and analytical insights throughout the project, tracked down and shook loose required data, and even rolled up his sleeves to help us do the "nug" work to extract performance data from complex and difficult-to-read charts and manuals.

In addition, many people throughout DoD provided great insight during this project, including Col David Tabor, Col David Clark of USAR TRADOC; Col Dan Grillone of SOCOM J31; Mark Beauchamp of USA INSCOM; Marvin Denenny of USA NSRWA; Daniel Turney of AFSOC/A5RV; LTC Stephen Loftis of USSOCOM; Kidd Manville of DSCA/STR/POL; CDR Brian Teets of JCS J5; John Thatch of PMA 226 NAVAIR; Col Bernard Willi of 438 AEW CAPTF; Capt Brent Golembiewski of ITAM-AF/A3, FOB Striker Baghdad; Lt Col Joseph Willoughby of OSD Policy; and Terry Pownall of USA ASA ALT.

We would very much like to thank those from the helicopter industry who not only provided aircraft performance and cost information, but also spent time discussing and helping us fully understand the details of their helicopters to ensure that we were modeling their aircraft correctly—from AgustaWestland: Cory Cave, Tracy Colburn, and John Lettieri; Bell Helicopter: Alan Ewing, Ted Trept, and Brad Wanek; Boeing: Bob Derham and John Thatch; EADS-NA: Stephen Mundt and Eric Walden; and Sikorsky: Steven Behnfeldt, Michael Pollack, Matthew Rodgers, and Dan Taylor.

Finally we would like to thank some of our RAND colleagues. Alex Hou, Shivan Sarin, and Brian Shannon, who worked on a different part of this project, provided help and insight to this work. David Frelinger and Dr. Edward Wu provided technical reviews of an earlier draft. Their reviews were very insightful and highly constructive. Jerry Sollinger, our communications analyst, helped us reorganize this document and improve the writing. We would also like to thank Holly Johnson, who helped prepare this manuscript.

Abbreviations

AFTOC	Air Force Total Ownership Costs
BW	basic weight
C	Celsius
CASEVAC	casualty evacuation
CER	cost-estimating relationship
DLR	depot-level repair
DoD	Department of Defense
DoS	Department of State
DV	distinguished visitor
FOB	forward operating base
FY	fiscal year
IGE	in-ground effect
ISA	International Standard Atmosphere
JOG	Joint Operations Graphic
LZ	landing zone
MEF	maximum elevation feature
MGTOW	maximum gross takeoff weight
MOB	main operating base
MoD	Ministry of Defense
MoI	Ministry of Interior
nm	nautical miles
NSRW	nonstandard rotary-wing
O&S	operations and support
OGE	out-of-ground effect
OSD-CAIG	Office of the Secretary of Defense Cost Analysis Improvement Group
OSMIS	Operating and Support Management Information System
SME	subject matter expert
STTO	startup, taxi, and take-off
USAF	U.S. Air Force

USG	U.S. government
VAMOSC	Visibility and Management of Operation and Support Cost

CHAPTER ONE

Introduction

The Department of Defense (DoD) provides significant assistance to partner nations in support of counterterrorism and counterinsurgency operations. Although one critical component of these operations is the partner nation's helicopter fleet, many of the United States' important security partners rely on equipment that is either not of U.S. origin or out of date. Indeed, over the last decade, DoD has increased security cooperation with a large number of states that fly such nonstandard rotary-wing (NSRW) aircraft. This fact presents particular challenges for the U.S. military, which seeks to conduct efficient partner support and security cooperation activities, but, for understandable reasons, does not regularly undertake the flight crew training, maintenance, or supply activities needed for the efficient operation of NSRW aircraft.

In a letter to the Secretary of Defense Senator Shelby points out that the United States has spent a considerable amount of money (over $800 million) purchasing new Mi-17s for Afghanistan and Iraq, and before the acquisition, (1) no requirements were defined, (2) no analysis of alternatives was completed, and (3) no other alternatives were considered.[1] Furthermore, multiple service program offices were involved in the program, and there were issues regarding the predictability of funds to support the acquisition.

Given U.S. investments in countries that rely upon NSRW aircraft, there is a need to consider whether increasing the U.S. capability to support these platforms is more cost-effective than aiding those partners in transitioning their fleets to different aircraft. This analysis, therefore, identifies and evaluates the cost-effectiveness of potential aircraft alternatives. This chapter provides an introduction to the methodology used to conduct analysis; we then describe each of the six components of the cost-effectiveness methodology and outline the structure of the rest of the report.

Six-Step Methodology

The methodology used to conduct this analysis comprises six steps:

1. delimiting the scope of cases to be included for analysis
2. identifying aircraft alternatives to be analyzed
3. developing what we call "mission-tasks"
4. evaluating helicopter effectiveness
5. estimating each helicopter's cost
6. aggregating the capability and cost to arrive at each helicopter's cost-effectiveness.

[1] Senator Richard Shelby, Letter to Secretary of Defense, October 21, 2009.

The first two steps are described here in full, while each subsequent step is briefly introduced before being explained in detail in the chapters that follow.

Definition of Analytical Cases

The first step of the RAND analysis was to set the scope of cases to be included for analysis. We have chosen to focus particularly on the helicopter requirements of four key partner nations—including Afghanistan and Iraq—along with 29 others.[2]

To make analysis of so many partners tractable, we differentiate the 29 countries into common groups based on mission distance and altitude. This produces a clustering around missions flown at low altitude for short and medium distances—less than 100 nautical miles (nm). We therefore focus our cost-effectiveness analysis on the two categories containing the largest fraction of the 29 nations evaluated: low-altitude/short-distance and low-altitude/medium-distance, and on the analytical boundaries of the mission space: high-altitude/short-distance; high-altitude/long-distance; low-altitude/long-distance. Each of these categories, with its constituent number of partner nations, is displayed in Table 1.1.

Definition of Helicopter Alternatives

The second step in the methodology is to identify the aircraft alternatives to be analyzed. We take the Mi-17v5 as the baseline aircraft for this analysis, with other helicopters considered as alternatives to acquiring additional Mi-17v5s for fleet expansion, or as candidates for replacing them as they are retired. We considered a number of military and civilian versions of utility (lift) and attack helicopters—analyzing such standard DoD aircraft as the Huey II, CH-47D Chinook, CH-46E Sea Knight, and UH-60 Black Hawks—but also several civilian models, including the AW109, the EC-145, and a number of Bell aircraft. In the attack category, we considered the Bell AH-1Z Viper and the AH-64D Apache, as well as the AW129 and the Mi-35. Several civil/military utility helicopters were also examined, the most notable being the Mi-17 and the S-61T. Table 1.2 displays the full set of alternative platforms included in our analysis.

Helicopters' *mission-capability* is determined by a number of design and environmental variables; for example, the altitude at which an aircraft must fly to clear obstacles, the nature of the terrain, and the temperature all affect the possible operating weight of the aircraft. The combination of these variables, along with sortie distance and hover performance, have impli-

Table 1.1
Grouping of Partner Nations for Analysis

		Typical Mission Distances			
		Short (10nm – 50nm)	Medium (50nm – 100nm)	Long (100nm – 200nm)	Total
Typical Landing Zone (LZ) Altitudes	Low (1,000' – 4,000')	6	10	4	20
	Medium (1,000' – 10,000')	3	1	1	5
	High (6,000' – 18,000')	1	1	2	4
	Total	10	12	7	

[2] Details on these partner nations are presented in a companion piece, not available to the general public: Christopher A. Mouton, David T. Orletsky, Michael Kennedy, Fred Timson, Adam Grissom, Akilah Wallace, *Cost-Effective Alternatives to the Mi-17 for Partner Nations: Focus on Afghanistan, Iraq, Pakistan, and Yemen*, Santa Monica, Calif.: RAND Corporation, 2014.

Table 1.2
Helicopter Alternatives Considered

Military Utility / Lift	Civilian Utility / Lift	Civilian / Military Utility / Lift	Attack
Huey II	Bell 407	S-92A	AH-1Z
UH-1N	Bell 429	AS-332L1	AH-46D
UH-1Y	Bell 412EP	S-61T	AW129
CH-46E	AW109	S-61T+[b]	Mi-35
CH-47D	EC-145[a]	AW139	
CH-47 Int		Mi-17v5	
AW149			
UH-60L			
UH-60M			
NH-90			
EC-725			
EH101			

[a] The LUH-72A Lakota is a variant of the EC-145.

[b] The S-61T+ is a new Sikorsky aircraft currently in the design phase that is designed to compete with the Mi-17v5. HIP-H is the older version of the Mi-17.

cations for payload, and even small changes in any one parameter may mean that a particular helicopter can no longer provide the needed payload and carry sufficient fuel to complete the sortie. As a result, several thousand performance values are required to model each alternative aircraft over a set of missions.

To produce robust and reliable results, we modeled each helicopter at a high level of detail. We rely upon three primary sources to ensure the depth and quality of the data used in the conduct of our analysis:

- For systems currently in U.S. military inventory, we use information found in performance manuals.
- For helicopters not currently in U.S. military inventory, we use information provided directly by the manufacturer.
- For helicopters of foreign origin, we relied on various other sources of data.
- These data were evaluated rigorously for consistency and accuracy.[3]

[3] Details on the helicopters included in our analysis are presented in Mouton et al., 2014, not available to the general public.

Development of Mission-Tasks

The third step in the cost-effectiveness methodology is to identify the capabilities that partners' helicopters need if they are to achieve U.S. security objectives. The purpose of U.S. assistance to these partner nation helicopter units is to enable them to conduct operations that further the interests of both themselves and the United States. We therefore relied upon U.S. military and DoD personnel to establish the set of necessary and sufficient performance standards. These are defined in terms of *mission-tasks,* a measure of what a helicopter fleet must be able to do. By connecting mission requirements to helicopter capability, mission-tasks link operational needs directly to fleet size for each helicopter analyzed.

Mission-tasks are not meant to be predictive, they are simply representations of the type of performance that partner fleets can be expected to need in the future. We therefore grounded our mission-tasks by referencing actual—that is, real-world—missions. As will be described in detail in Chapter Two, we use real-world information to help us define mission-tasks such that they determine the total amount of personnel and equipment to be moved, the time frame of the movement (e.g., single lift, one period of daylight), the departure base, and the landing zone, but they do not dictate the number of passengers on each sortie.[4]

Mission-tasks also include a description of the temperature conditions under which the movement must take place, a variable that can influence greatly the performance of each helicopter. Together, these parameters produce a set of rotary requirements that are country-specific and include a wide range of missions, including movement of forces, sustainment of forces, medical evacuation, and casualty evacuation (CASEVAC).

Effectiveness Analysis

With the mission-tasks thus defined, the fourth step in the cost-effectiveness methodology is to analyze each helicopter's effectiveness. To arrive at a meaningful measure of each alternative's effectiveness relative to the Mi-17v5, we analyzed multiple mission-tasks on each of the 100 routes, at each of three climate conditions, for a total of 338 effectiveness values per alternative helicopter, per temperature—that is, more than 1,000 effectiveness values for each helicopter. To manage this level of complexity, we designed and applied a model with automated features capable of evaluating multiple functional and environmental parameters simultaneously. The model's output is the number of aircraft needed to complete the given mission-task, a figure that informs each helicopter's effectiveness score.

The result of this process, which is explained in full in Chapter Three, is an *effectiveness score,* which is the ratio between the number of Mi-17v5s and the number of the alternative aircraft required to carry out the given mission-task; i.e., quantity of Mi-17v5s divided by quantity of the alternative. This metric considers only those mission-tasks that the Mi-17v5 can perform and excludes helicopters that cannot do those same missions; however, we perform a sensitivity analysis to ensure that there are no helicopters that would otherwise be cost-effective. Thus, if the number of alternative aircraft required to do a set of tasks is the same as the number of Mi-17v5s needed, the effectiveness score of the alternative is 1.0—the two are

---

[4] Defining the minimum number of passengers on a sortie should be avoided, unless there is an operationally significant reason to do so, because it can cause alternatives to fail and be eliminated from further consideration. For example, if the task is to deploy a company of 100 personnel in a single lift, is it critical to have 20 personnel per aircraft (resulting in five sorties) or would 15 passengers per aircraft be acceptable? It is clear there is some minimum number of passengers that is operationally significant, and minimums should be established to eliminate completely unreasonable cases. It is best, however, to set this limit low so that the most alternatives are carried through the analysis.

equivalent in military capability. But if the number of the alternative craft required to do a set of tasks is half the number of Mi-17v5s needed, the effectiveness score of the alternative is 2.0; i.e., it has twice the military capability of an Mi-7v5, and so forth (Table 1.3).

Cost Analysis

The fifth step in the RAND analysis is to measure cost, again relative to the Mi-17v5. Because the aircraft analyzed are existing designs currently in production, we have excluded from our analysis the costs of development that usually would be included in an aircraft's "cradle-to-grave" life cycle. The total life-cycle cost of an aircraft depends on both the flyaway cost of the aircraft and its operations and support (O&S) cost.

We obtained flyaway cost data from various outlets, including the U.S. government (USG), contractors, and trade press and other open-source media. For the Mi-17v5, we relied upon recent contracts and databases maintained by the USG, as well as two USG reports.[5] We requested price information from all of the contractors producing the helicopter alternatives included in our analysis, and received responses from Bell, Boeing, Sikorsky, and Eurocopter. Where available, we accepted the valuation of the helicopters' commercial variants as provided by aviation consultancy Conklin & de Decker. In cases where even this information was unavailable, we used a "comparable data" approach; we were unable to arrive through any of these methods at a reliable estimate for three helicopters, the AW129, the Mi-35, and the Mi-17 (HIP-H). USG data were included in the model for all aircraft for which they were available, unless the contractor-provided estimate was higher.[6]

While sources provided generally consistent data on procurement price, the same cannot be said for O&S costs. To the contrary, there is great variation in how dollar-per-flying-hour O&S estimates are calculated. Although some data repositories (for example, Conklin & de Decker) do provide consistent estimates of O&S costs for a number of platforms, their accounting does not include the full set of helicopter alternatives being analyzed here. We therefore use the May 1, 1992, Office of the Secretary of Defense Cost Analysis Improvement Group (OSD-CAIG) structure to define the categories of O&S to be included in our model.[7] We take as the basis for

Table 1.3
Effectiveness Criteria

Relative Capability	Effectiveness Score
10 alternatives have the mission capability of 20 Mi-17v5s	2.0
20 alternatives have the mission capability of 20 Mi-17v5s	1.0
40 alternatives have the mission capability of 20 Mi-17v5s	0.5

[5] DoD, *Mi-17 Helicopters,* March 23, 2010b; Department of State (DoS), *Aviation Services Study,* March 9, 2009.

[6] We saw no evidence to suggest contractor prices were biased, either up or down.

[7] This structure is consistent with that of the U.S. Air Force Total Ownership Cost (AFTOC). Office of the Secretary of Defense Cost Analysis Improvement Group, *Operating and Support Cost-Estimating Guide,* May 1, 1992.

our model the AFTOC structure, supplementing with Army and Navy/Marine Corps helicopter maintenance data when available. This structure is designed to capture the characteristics of the helicopters in use by the U.S. military; as such, we did not assume it would be equally applicable to the Mil Moscow Helicopters included in the RAND analysis, the Mi-17v5 and the Mi-35.

Just as with effectiveness, we operationalized cost as a ratio between the baseline Mi-17v5 and the alternative platform, using the unit cost of the Mi-17v5 and the unit cost of the alternative helicopter. If the cost of the alternative is the same as that of the Mi-17v5, its cost value is defined as 1.0. If its cost is twice that of the Mi-17v5, its cost value is defined as 2.0, and so on. As will be discussed in Chapter Four, we use the term *cost* to mean the net present value of all life-cycle expenditures, including acquisition, training, maintenance, and military construction costs.

Integration of Effectiveness and Cost Analysis

The sixth and final step of our analysis is to arrive at each helicopter's cost-effectiveness as compared to that of the Mi-17v5. We define *cost-effectiveness* as the ratio of the alternative's effectiveness score to its cost score. The meaning of this measure is illustrated in Table 1.4. If the alternative has half the cost and half the military effectiveness of the Mi-17v5, *or* if it has twice the cost and twice the military effectiveness of the Mi-17v5, then its cost-effectiveness is 1.0. Spending a given budget on Mi-17v5s or on the alternative gives the same level of military capability, and they are equally good in these terms. If an alternative has the same capability as the Mi-17v5 and costs half as much, or if it has twice the capability of the Mi-17v5 and costs the same, its cost-effectiveness is 2.0; i.e., spending a given budget on the alternative gives twice the military capability as spending it on the Mi-17v5. The cost-effectiveness ratios continue in this pattern (superior alternatives are in blue cells, and neutral options are green).

Report Structure

Consistent with the six-step methodology described, Chapters Two through Five compose the analytical heart of the report. Chapter Two presents the method by which mission-tasks and the associated routes were developed. In Chapter Three, we explain how these mission-tasks were applied in the evaluation of helicopter performance; we describe the model designed to quantify each platform's effectiveness, and present its results. Chapter Four presents our cost analysis in full, and Chapter Five integrates the effectiveness results with the cost estimates to arrive at concrete measures of cost-effectiveness. We present our conclusions in Chapter Six.

Table 1.4
Relative Cost and Effectiveness

		Relative Effectiveness		
		0.5	1.0	2.0
Relative Cost	0.5	1.00	2.00	4.00
	1.0	0.50	1.00	2.00
	2.0	0.25	0.50	1.00

CHAPTER TWO

Mission-Task Development

This chapter addresses the third step in the cost-effectiveness methodology: identifying the capabilities that partner helicopters need if they are to achieve U.S. security objectives. We begin by describing how we identified appropriate performance standards and operationalized them in terms of what we call *mission-tasks.* These mission-tasks, in turn, informed the definition of the routes that helicopters must be able to fly. This process is unique to each environment, and so is explained within the context of each key partner nation examined. We similarly describe the method for developing a tractable number of mission-tasks and associated routes for the 29 other partner nations included in the set of countries analyzed.

Defining Mission-Tasks

The purpose of U.S. assistance to these partner nation helicopter units is to enable them to conduct operations that further the interests of both the partner country and the United States. We therefore relied upon U.S. military and DoD personnel to establish the set of necessary and sufficient performance standards. These are defined in terms of *mission-tasks,* a measure of what a helicopter fleet must be able to do. By connecting mission requirements to helicopter capability, mission-tasks link operational needs directly to fleet size for each helicopter analyzed.

We defined mission-tasks such that they determine the total amount of personnel and equipment to be moved, the time frame of the movement (e.g., single lift, one period of daylight), the departure base, and the landing zone, but they do not dictate the number of passengers on each sortie. These characteristics allowed us to calculate mission distance, and to account for en route terrain and altitude. Together, these data produce a set of helicopter requirements that are country-specific and include a wide range of missions, such as movement of forces, sustainment of forces, medical evacuation, and CASEVAC.

Mission-tasks can be additive, or the maximum of a set of mission-tasks may be the important consideration. For example, the requirement used for analysis might be the most demanding of either the movement of a battalion from a base to landing zones (LZs) or the movement of several companies to different LZs simultaneously. Mission-tasks also include a description of the temperature conditions under which the movement must take place, a variable that can greatly influence the performance of each helicopter.

Mission-tasks are not meant to be predictive, but rather to represent the type of performance that partner fleets can be expected to need in the future. We therefore grounded our mission-tasks by referencing actual—that is, real-world—missions. For example, after the May 2010 crash of Pamir Airways Flight 112 in Afghanistan, an International Security Assistance Force helicopter delivered a rescue force of 200 mountain climbers to the crash site. This operation is not modeled in all its specifics in our analysis, but our mission-tasks do incorporate a number of its characteristics: One follows its flight path, another applies a similar mission radius, and a third echoes its infiltration size. This mission-task approach allowed us to gain insight into rotary-wing aircraft that are suitable for different types of missions without requiring that we specifically model each and every potential such mission. That is, we were able to define reasonable proxies for future missions that span different magnitudes of mission characteristics.

Figure 2.1 presents this approach graphically. Parameters A and B can be any of the aspects that define the mission-tasks. For example, Parameter A might be mission distance, while Parameter B might be LZ altitude. A set of mission-tasks, represented by the nodes on the lattice network in the diagram, encompass and bound the mission sets, enabling the methodology to identify helicopter performance capabilities across the spectrum of possible missions.

To develop mission-tasks, we used prior RAND research that describes the types of operations being conducted, the bases being used, and the regions in which they occur; we then refined this general understanding for our key partner nations through discussions with subject matter experts (SMEs) and experts on the countries in question.[1] Using this information, we designed a range of mission-tasks with variations on initiating base, LZ, range, en route terrain features, operational activity (e.g., infiltrating a company during a period of daylight), and other relevant characteristics. With the mission-tasks thus defined, we next turned to developing country-specific flight routes; we will describe this process in more detail.

Figure 2.1
Spectrum of Mission-Tasks

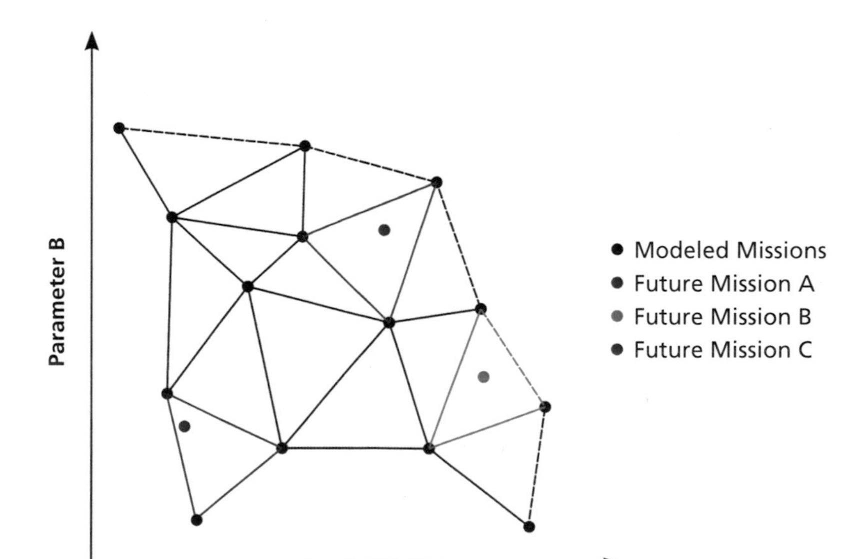

RAND *RR141z1-2.1*

[1] Mission-task details are presented in Mouton et al., 2014, not available to the general public.

Defining Routes

Mission-tasks cannot provide a basis for evaluating aircraft capability in the absence of defined transit routes. It is these that set values for two variables that affect fundamental helicopter performance: range, and the altitudes of bases, LZs, and terrain features. Starting with the area of operations and primary bases designated in the mission-tasks, we next selected a primary LZ for each mission-task in each partner nation (see Figure 2.2). Other LZs were then dispersed a constant distance around the primary one in different directions (angles). When this vector located an LZ across an international border, we rotated the angle or reduced the range to return the LZ to the correct side of the line. In other cases, we provided a main operating base (MOB) or forward operating base (FOB) from which operations would be conducted. In these instances, the operations were not located in a single area of responsibility, but rather were more dispersed; for example, see Afghanistan Ministry of Defense (MoD) missions (the right diagram in the figure). Locations for LZs were then identified by projecting vectors at equal increments (120 degrees in Figure 2.2), again rotating or reducing range when a border was crossed, and locating the most proximate town.

Once the LZs were identified, we could then develop multiple flight paths for each mission. Our approach was to begin with a direct route between the initiating base and the mission-identified LZ. If this required flying over a significant terrain feature that could be avoided by choosing a different route, we developed alternative routes for the base-LZ pair that were longer, but allowed flight at lower altitude. To do this, we worked with topographic maps, primarily Joint Operations Graphics (JOGs), which have a high degree of fidelity, to determine required distance and altitude. The maximum elevation feature (MEF) of each quadrangle allowed us to establish the minimum altitude required for each flight path.

Upon completion, all mission-tasks were reviewed by U.S. personnel in-theater or by a country expert. The result is a sufficiently sized set of mission-tasks that are reliable approxima-

Figure 2.2
Route Definition

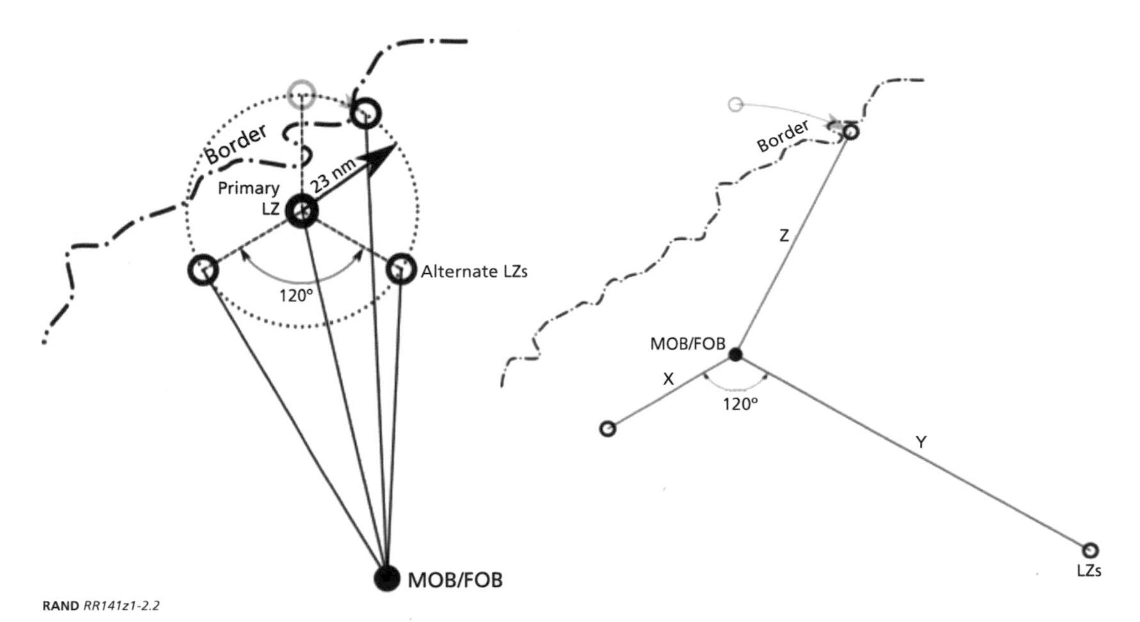

tions of the types of missions that rotary-wing aircraft can be expected to be conducted. The mission-tasks developed are specific to each country and are described in detail in the following subsections.

Afghanistan Mission-Tasks

We identified three sets of missions for Afghanistan: for the MoD; the Ministry of Interior (MoI); and distinguished visitors (DVs).

Afghanistan MoD Mission Characteristics

Based on conversations and interaction with U.S. personnel in-theater, we identified three Afghan MoD missions that helicopters must be able to perform: troop infiltration, force sustainment and CASEVAC, and attack.[2]

We sized two mission-tasks for troop infiltration operations: a company-sized force and a battalion-sized force. The company-sized mission-task was defined as a single-lift movement. For the battalion-sized mission-tasks, we designed both single- and multiple-sortie options, with the latter occurring during an 11-hour period of daylight. For both the company and the battalion mission-tasks, the average per-troop weight (including personal equipment and one day's worth of supplies) was set at 220 pounds based on discussions with U.S. planners in-country.

We similarly sized two daylight mission-tasks for the sustainment/CASEVAC operation, again for a company and for a battalion. We assumed a need for 40 pounds of cargo per troop, per day, and a retrograde CASEVAC requirement of 10 percent for the smaller company-sized force and 5 percent for the battalion-sized force.

There are three types of attack missions included in the analysis. The first is fire, or close air support, and requires attack helicopters to be airborne and on-call for six hours. The second mission is battlefield preparation, in which a single sortie provides fire prior to a troop infiltration operation. The third and final attack mission is escort. Here, the requirement is for the attack helicopters to accompany utility aircraft during a single-sortie infiltration operation.

Using the process described, 20 mission-task routes were defined for Afghanistan MoD mission-tasks. These routes all use six regimental bases, each with three associated landing zones, displayed in Figure 2.3.

Table 2.1 displays the name and altitude of each base and LZ, the distance from base to LZ, maximum en route altitude, and mission radius. Routes 7 and 12 have direct great-circle routes and feasible alternate routes that avoid high-terrain obstacles, and these are indicated as routes a and b, respectively; the other 16 missions either cross terrain low enough that an alternate route would not be advantageous or have no feasible alternative.

Figure 2.4 provides a graphic representation of the diversity and difficulty of the sets of routes included for analysis, displaying simultaneously mission radius (x-axis) and altitude (y-axis) for each route's landing zone and highest terrain feature. For example, Route 1 has a mission radius of 30 nm, an LZ altitude of 1,000 feet as indicated by a red square, and a maximum en route altitude of 1,300 feet, as indicated by the blue diamond above the red square.

---

[2] Further details on these missions are presented in Mouton et al., 2014, not available to the general public.

Figure 2.3
Afghanistan MoD Mission-Task Routes

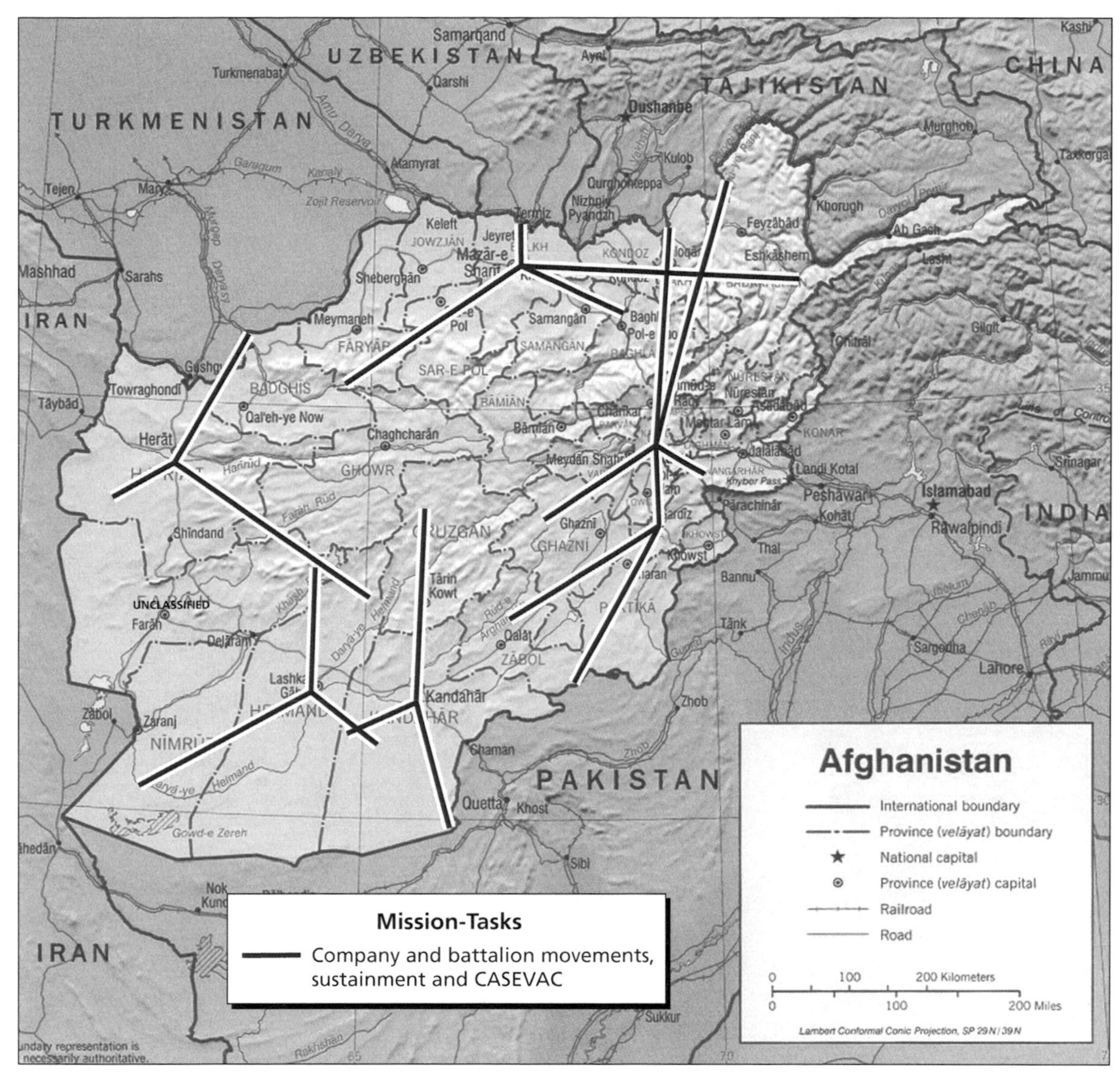

RAND *RR141z1-2.3*

Table 2.1
Afghanistan MoD Mission-Task Route Information

Route Number	Base	Landing Zone	Base Altitude (ft)	Landing Zone Altitude (ft)	Maximum En Route Altitude (ft)	Mission Radius (nm)
1	Mazari Sharif	Hazareh Toghay	1,300	1,000	1,300	30
2	Mazari Sharif	Baghlan	1,300	1,700	8,400	81
3	Mazari Sharif	Takhaspun	1,300	6,100	12,000	154
4	Mazari Sharif	Zibak	1,300	8,700	17,000	201
5	Kabul	Dasht-e Qal'eh	5,900	1,700	16,500	157
6	Kabul	Neyan Khwar	5,900	6,600	12,400	42
7a	Kabul	Kuh-e Budak	5,900	13,700	15,400	98
7b	Kabul	Kuh-e Budak	5,900	13,700	13,700	111
8	Kabul	Cosnukel	5,900	4,500	17,200	199
9	Gardez	Kabul	7,800	5,900	11,900	56
10	Gardez	Serkey Kalay	7,800	6,900	12,500	130
11	Gardez	Kuchnay Zardalu	7,800	6,600	10,100	130
12a	Kandahar	Khakrez	3,300	9,900	13,600	138
12b	Kandahar	Khakrez	3,300	9,900	12,500	145
13	Kandahar	Nalai Narai	3,300	5,800	6,700	96
14	Kandahar	Kuchnay Wastah	3,300	3,600	4,800	57
15	Lashkar Gah	Pire Kermah	2,500	9,400	10,600	90
16	Lashkar Gah	Fatehnawer	2,500	3,800	4,700	62
17	Lashkar Gah	Bandare Wasate	2,500	1,700	3,600	150
18	Herat	Muricag	3,200	1,500	9,600	115
19	Herat	Kelak	3,200	10,900	14,300	149
20	Herat	Chāh-e Gazak	3,200	3,600	6,600	53

Figure 2.4
Afghanistan MoD Mission-Task Radius and Altitude

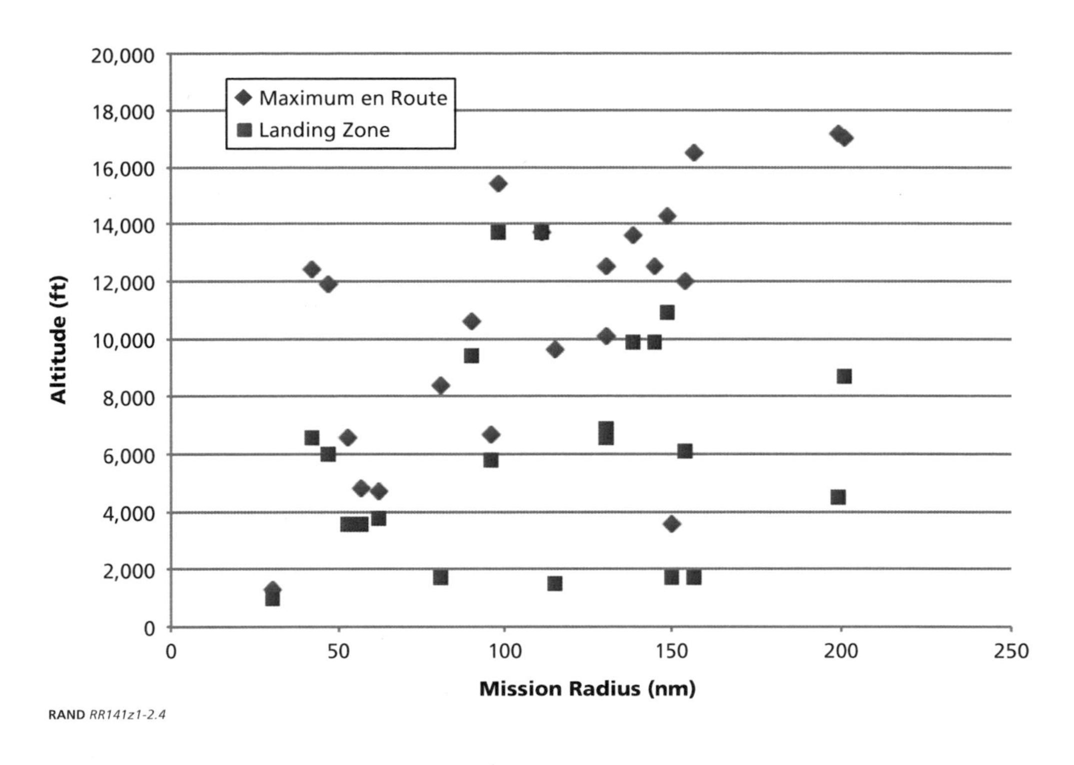

RAND *RR141z1-2.4*

Afghanistan MoI Mission Characteristics

Two missions produced mission-tasks for the Afghanistan MoI, the first of which is an assault mission. This entails insertion of an assault in a single lift, a 15-minute stay in the area, and then troop extraction. For these missions, we identified four operational bases, which produced mission distances of 25 to 76 nm. The second MoI mission-task is resupply, which requires swapping troops and transporting replacement supplies for an outpost. This mission was sized based on discussions with U.S. planners in-theater. Figure 2.5 displays the 19 mission-task routes developed for these MoI missions: the blue lines denote assault; the red lines, operational resupply.

Table 2.2 displays the name and altitude of each base and LZ; the distance from base to LZ; maximum en route altitude; and mission radius. Here, again, assault missions are in blue, and resupply in red.

Figures 2.6 and 2.7, respectively, display the variety of routes included in MoI assault and resupply mission-tasks, presenting graphical representation of mission radius, LZ altitude, and maximum en route altitude.

Afghanistan DV Mission Characteristics

All representative DV routes originate in Kabul and terminate in one of Afghanistan's major cities: Jalalabad, Kunduz, Mazari Sharif, Taluqan, Khost, Kandahar, Lashkar Gah, or Herat (Figure 2.8). The missions are flown from airbase to airbase, permitting in-ground effect hover and enabling the helicopters to refuel at the outbound destination. DV routes are longer than those designed to represent the needs of the MoD and the MoI, but they include a refueling option and have relatively lighter payloads.

Figure 2.5
Afghanistan MoI Mission-Task Routes

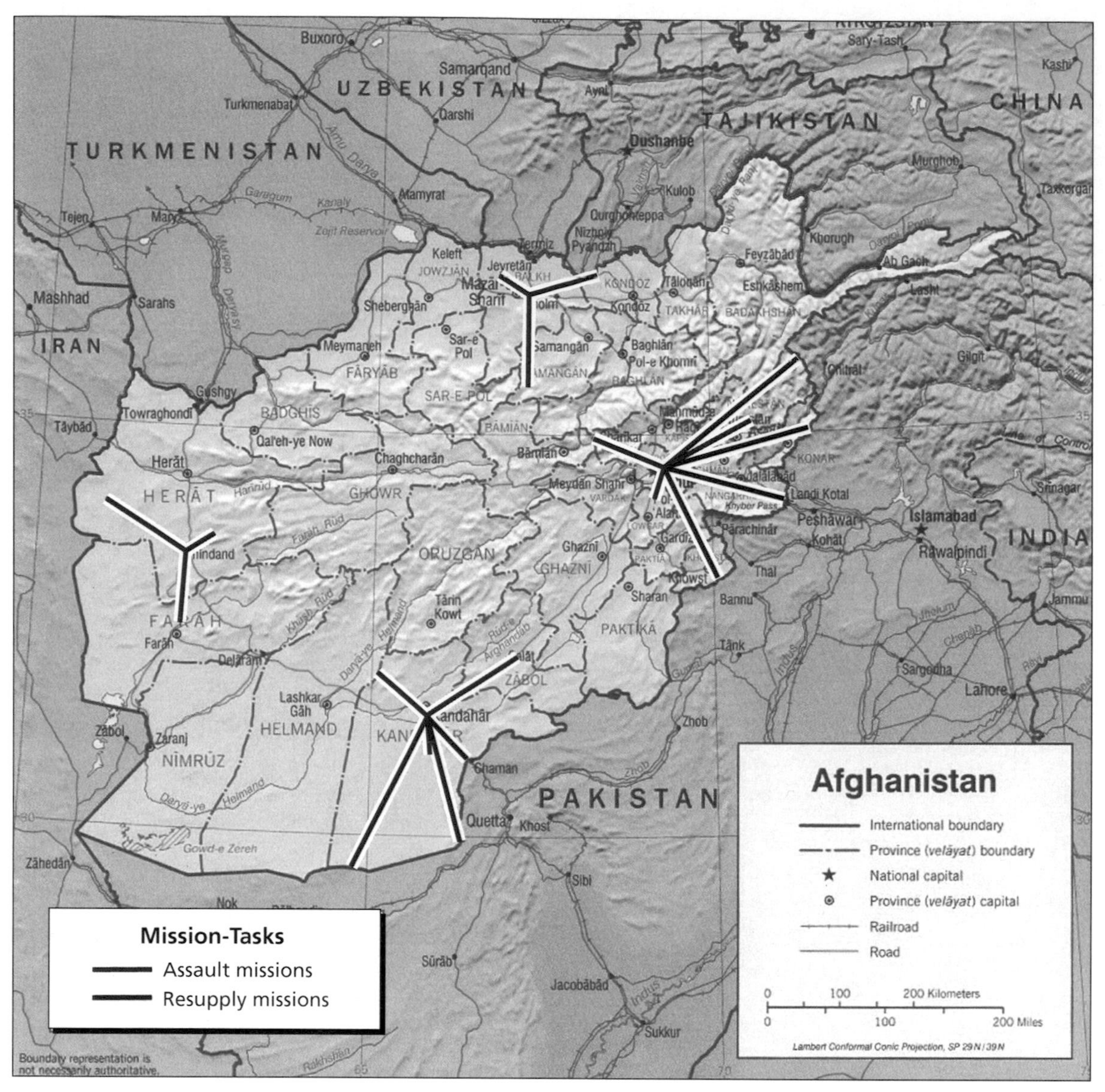

RAND *RR141z1-2.5*

Table 2.2
Afghanistan MoI Mission-Task Route Information

Route Number	Base	Landing Zone	Base Altitude (ft)	Landing Zone Altitude (ft)	Maximum En Route Altitude (ft)	Mission Radius (nm)
1	Kabul	Diwanah Baba	5,900	9,100	12,400	168
2	Kabul	Zor Barawui	5,900	7,100	11,100	118
3	Kabul	A-1 Border Cross	5,900	2,800	11,600	97
4	Kabul	Golamkhan Kalay	5,900	4,400	14,700	95
5	Kandahar	Tut Kalay/A-75	3,300	4,200	6,100	46
6	Kandahar	Nalai Narai	3,300	5,800	6,700	96
7	Kandahar	Derangi Chapar	3,300	2,900	5,700	128
8	Kandahar	Tangay Ana Ziarat	3,300	4,100	6,300	28
9	Kandahar	Khele Jhulaman	3,300	5,500	8,200	76
10	Kandahar	Cenar	3,300	3,700	9,300	49
11	Kabul	Moghulkhel	5,900	6,100	10,300	25
12	Kabul	Sar-e-Kowtal	5,900	9,000	15,900	53
13	Kabul	Gulcha	5,900	8,000	15,300	75
14	Mazari Sharif	Soltan Hajji Wali	1,300	1,000	1,500	25
15	Mazari Sharif	Baghri Kol	1,300	1,100	2,000	53
16	Mazari Sharif	Safedkhak	1,300	7,700	13,600	72
17	Shindand	Sach	3,800	2,300	7,700	52
18	Shindand	Cah-I-Rabat	3,800	2,300	6,300	72
19	Shindand	Mirabad	3,800	6,600	10,300	25

NOTE: Assault missions are signified with blue cells, resupply missions are in red.

Figure 2.6
Afghanistan MoI Assault Mission-Task Radius and Altitude

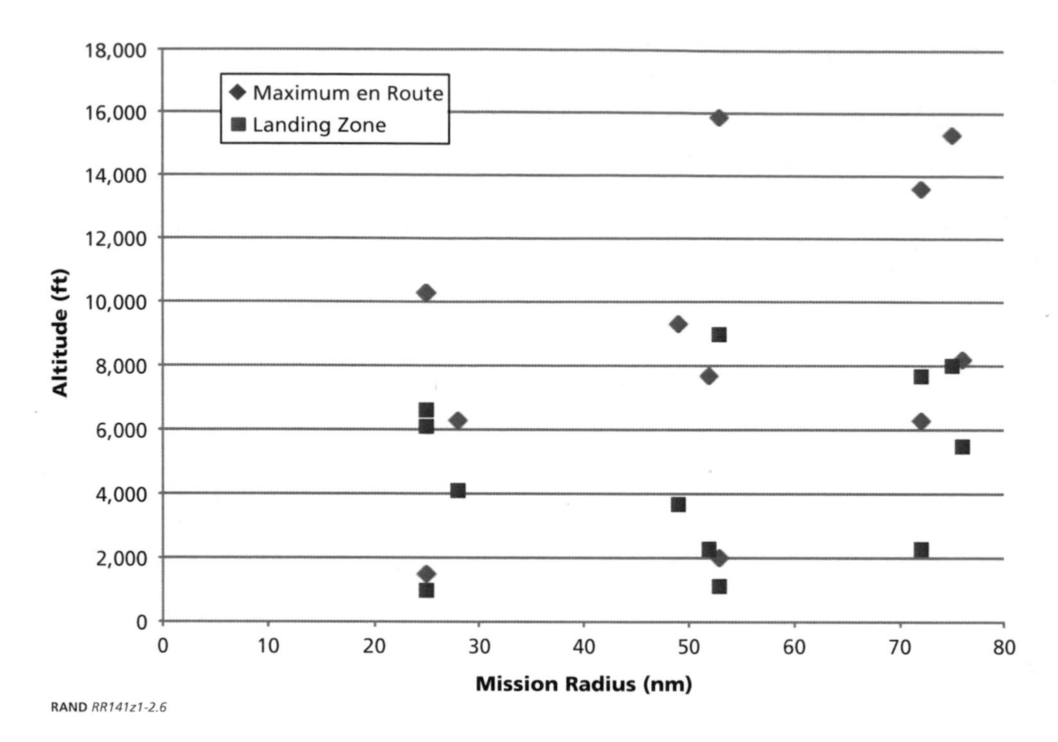

RAND *RR141z1-2.6*

Figure 2.7
Afghanistan MoI Resupply Mission-Task Radius and Altitude

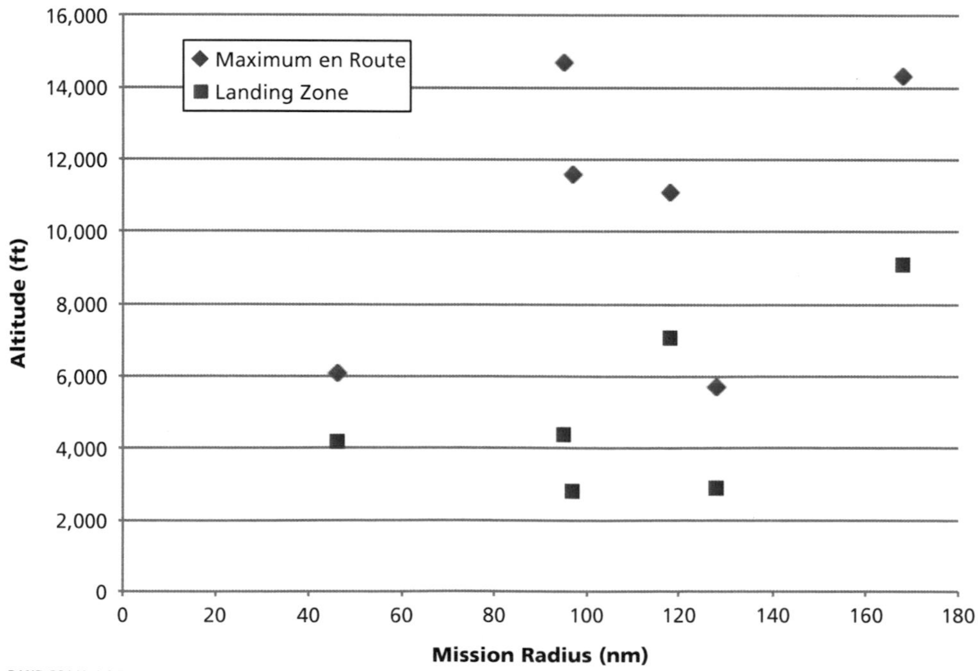

RAND *RR141z1-2.7*

Figure 2.8
Afghanistan DV Mission-Task Routes

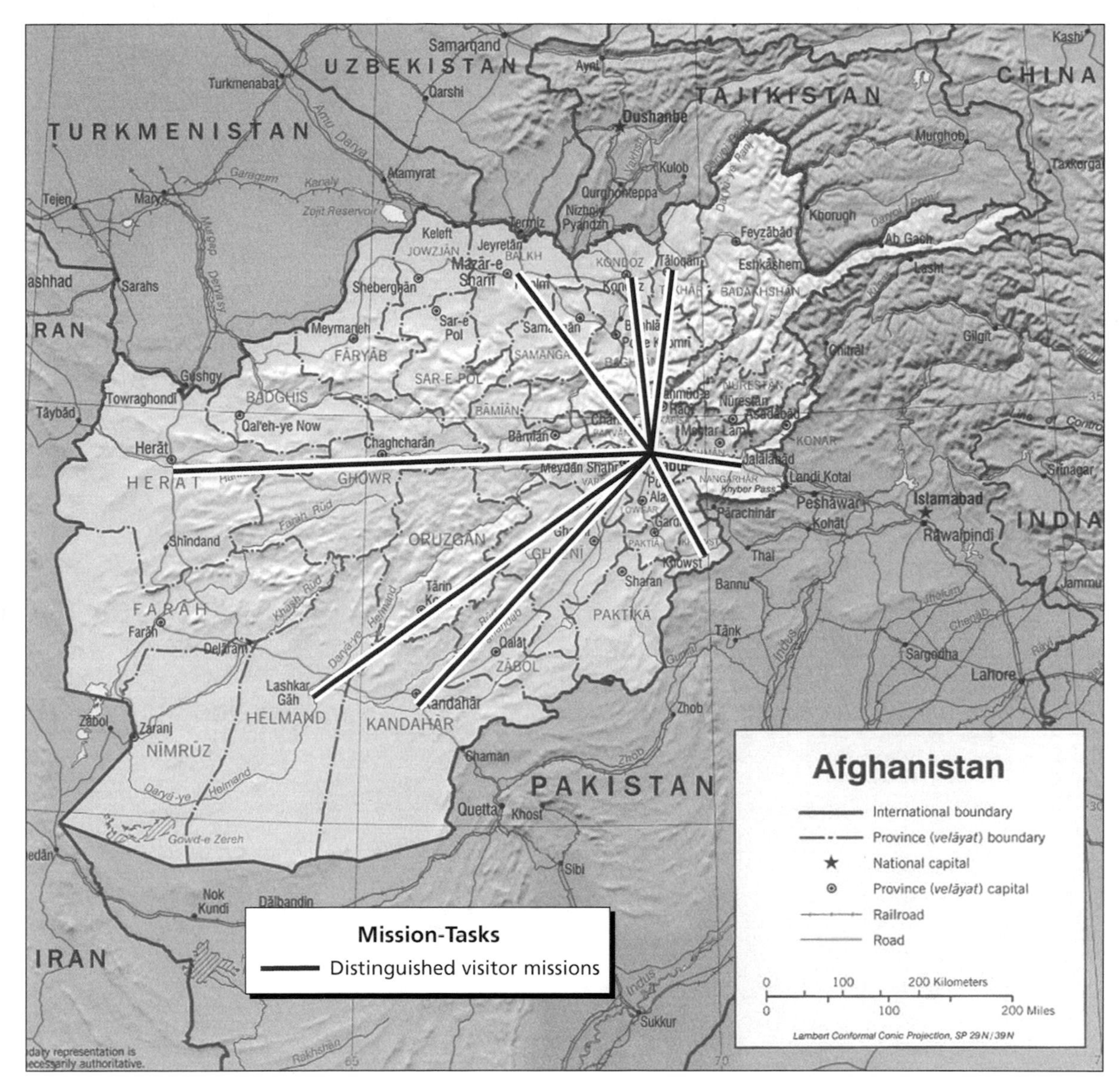

RAND *RR141z1-2.8*

Additional information about these routes is provided in Table 2.3; note that Route 5 has an alternative that avoids high-terrain obstacles.

Finally, Figure 2.9 relates mission radius to altitude requirements. Of note here is that while the LZ altitudes for DV missions are fairly low, many routes traverse high-terrain features.

Iraq Mission-Tasks

The mission-tasks for Iraq are similar to the infiltration, sustainment/CASEVAC, and attack missions defined for the Afghanistan MoD. Operations originate from five major bases: Taji, Al Taquddum, Al Kut, Basra, and Kirkuk. Company movements required single-lift, while

Table 2.3
Afghanistan DV Mission-Task Route Information

Route Number	Base	Landing Zone	Base Altitude (ft)	Landing Zone Altitude (ft)	Maximum En Route Altitude (ft)	Mission Radius (nm)
1	Kabul	Jalabad	5,900	1,800	9,900	65
2	Kabul	Kunduz	5,900	1,400	16,200	136
3	Kabul	Mazari Sharif	5,900	1,300	16,200	182
4	Kabul	Taluqan	5,900	2,700	16,299	136
5a	Kabul	Khost	5,900	3,800	14,700	83
5b	Kabul	Khost	5,900	3,800	12,400	95
6	Kabul	Kandahar	5,900	3,300	12,300	258
7	Kabul	Lashkar Gah	5,900	6,400	12,300	326
8	Kabul	Haret	5,900	3,200	14,500	358

Figure 2.9
Afghanistan DV Mission-Task Radius and Altitude

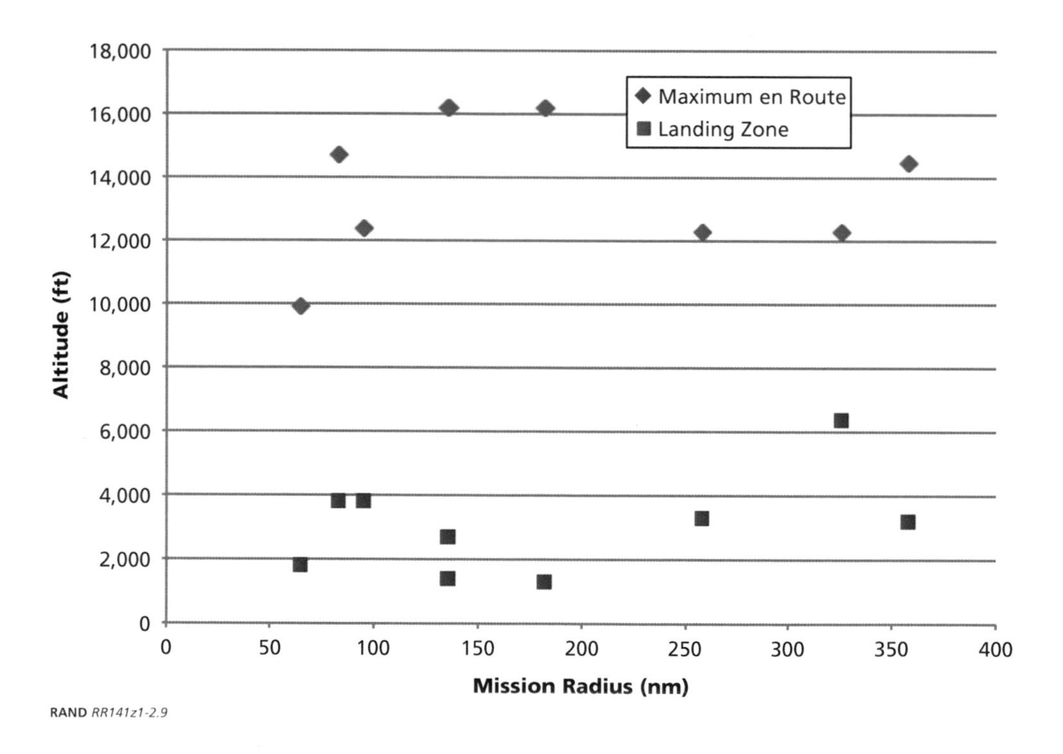

RAND *RR141z1-2.9*

battalions operations were given an 11-hour period, with the assumed weight per passenger set at 220 pounds. The set of 17 Iraq mission-task routes are displayed in Figure 2.10.

Table 2.4 provides additional detail about the 17 representative routes developed for Iraq. They are notable for their relatively low altitudes.

These altitudes and distances are represented graphically in Figure 2.11.

Figure 2.10
Iraq Mission-Task Routes

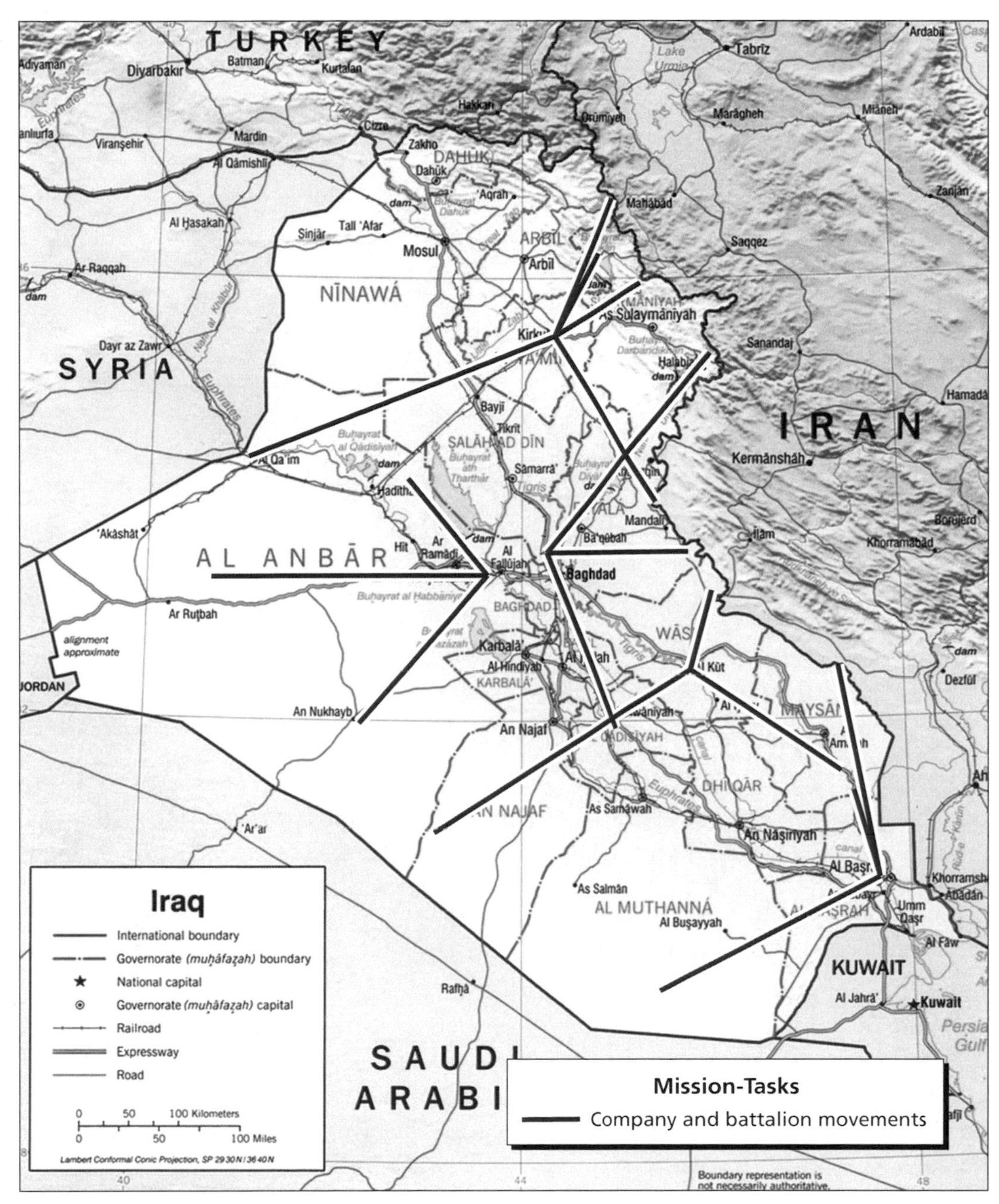

RAND *RR141z1-2.10*

Table 2.4
Iraq Mission-Task Route Information

Route Number	Base	Landing Zone	Base Altitude (ft)	Landing Zone Altitude (ft)	Maximum En Route Altitude (ft)	Mission Radius (nm)
1	Taji	Khurmal	100	1,800	7,600	139
2	Taji	Ad Diwaniyah	100	500	1,300	100
3	Taji	Salman Farraj	100	300	1,400	74
4	Al Taquddum	An Nukhayb	300	1,000	1,300	104
5	Al Taquddum	Mileh Tharthar	300	600	1,200	66
6	Al Taquddum	Pumping station	300	1,800	2,600	146
7	Al Kut	Imam Rada	100	300	1,100	43
8	Al Kut	Qal'a Salih	100	0	600	97
9	Al Kut	Al Ashuriyah	100	1,000	1,500	160
10	Basra	Banilam Region	0	700	1,200	119
11	Basra	Al'Uzay	0	0	500	49
12	Basra	Jahaym	0	1,000	1,400	130
13	Kirkuk	Husaiba	1,200	600	2,000	175
14	Kirkuk	Buhayrat'Dokan	1,200	1,700	5,500	50
15	Kirkuk	Makatu	1,200	600	3,600	109
16	Kirkuk	Galalah	1,200	3,600	9,300	56
17	Kirkuk	Ziriu	1,200	6,200	12,000	79

Mission-Tasks: 29 Partner Nations

To make analysis of the 29 additional partner nations tractable, we differentiated them into groups of common mission range and altitude. To achieve this, we used U.S. security interests to identify the countries' key operational areas and proximate actual and/or potential air facilities. This provided us a means of calculating the types of distances and altitudes—of bases, LZs, and en route terrain features—that helicopters must be able to traverse to be effective.

Figure 2.12 displays the highest altitude required and the maximum mission radius for each of the 29 partner nations. Because these data points represent the upper boundary required of helicopter performance, mission-tasks designed to meet these thresholds will capture all other altitudes and radii as lesser, included cases.

Notable in the figure is the clustering of partner nations into what we call "baskets," at low altitude for short and medium distances (less than 100 nm): low-altitude/short-distance and low-altitude/medium-distance. In this report we will focus on the boundaries of the analytical space: low-altitude/short distance; high-altitude/short-distance; high-altitude/long-distance; low-altitude/long-distance.[3]

For each of the altitude-range categories, we use an archetype nation to develop realistic mission-tasks that are applicable to all countries in the basket, without being specific to any one. The radius and altitude boundaries that delimit each archetype are displayed in Figure 2.13.

[3] Detailed results for all the country baskets are presented in Mouton et al., 2014, not available to the general public.

Figure 2.11
Iraq Mission-Task Radius and Altitudes

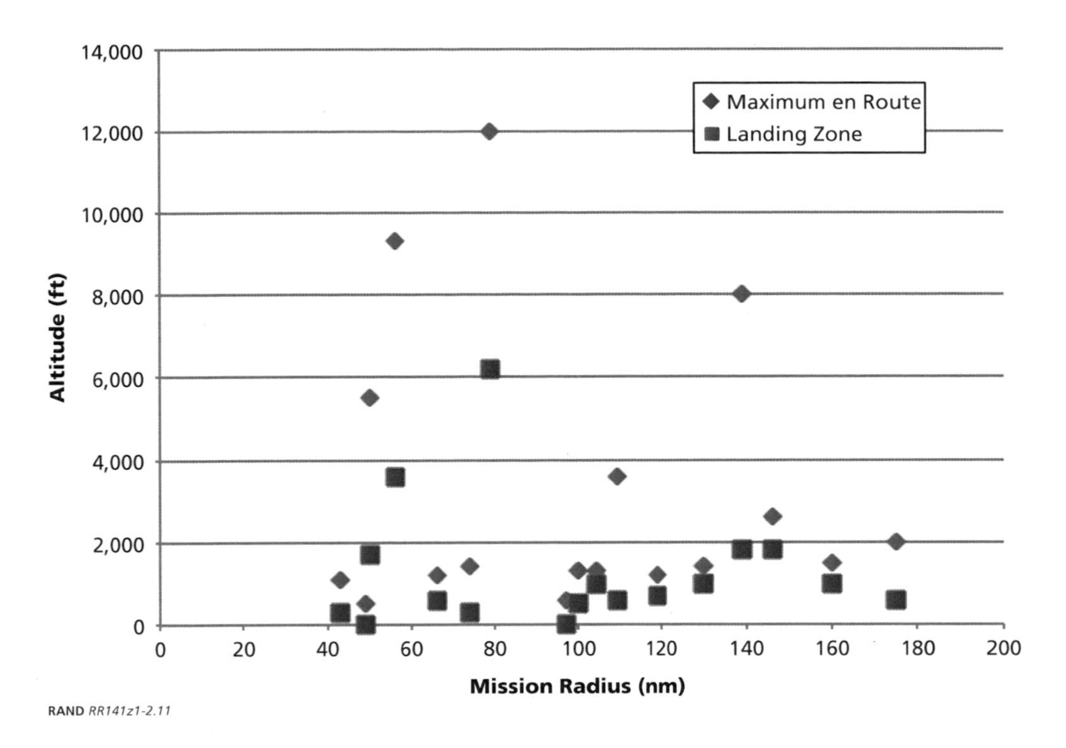

RAND *RR141z1-2.11*

For each of these nine archetype partner nations, we then developed mission-tasks and representative routes. We developed mission-tasks in key operational areas and varied LZ locations within those areas to expand the number of mission-tasks in each of the archetype counties, thereby getting a larger set of mission-tasks.

Figure 2.12
Partner Nation Mission-Task Radius and Altitude

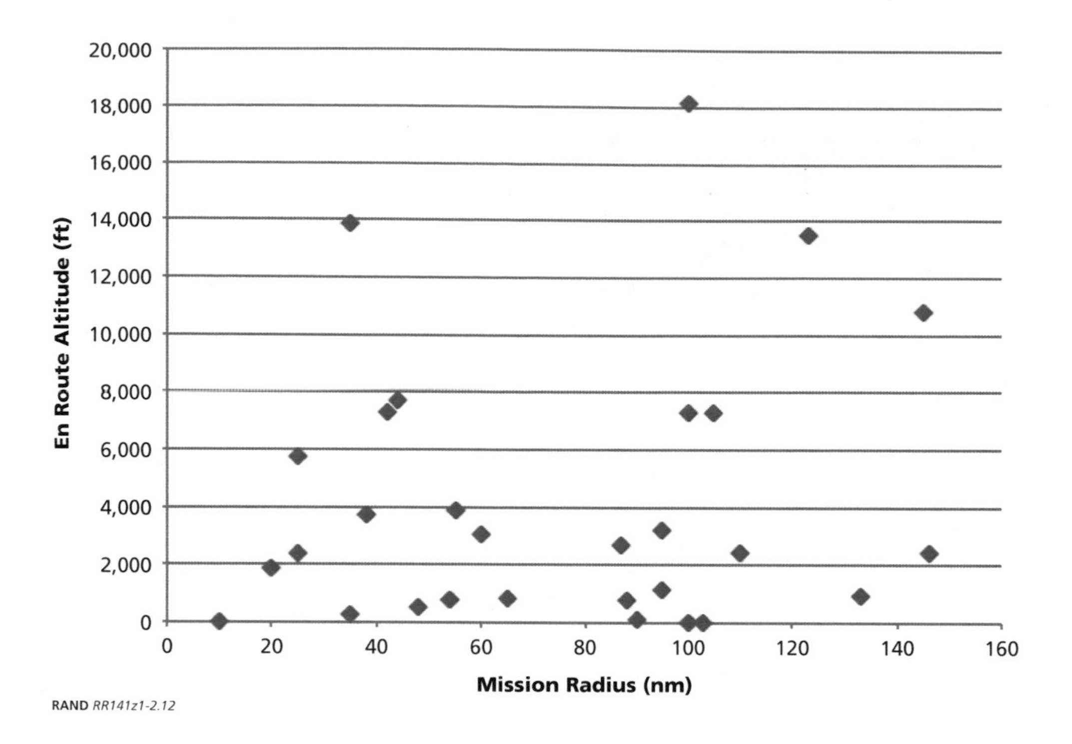

RAND *RR141z1-2.12*

Figure 2.13
Archetype Partner Nation Definitions

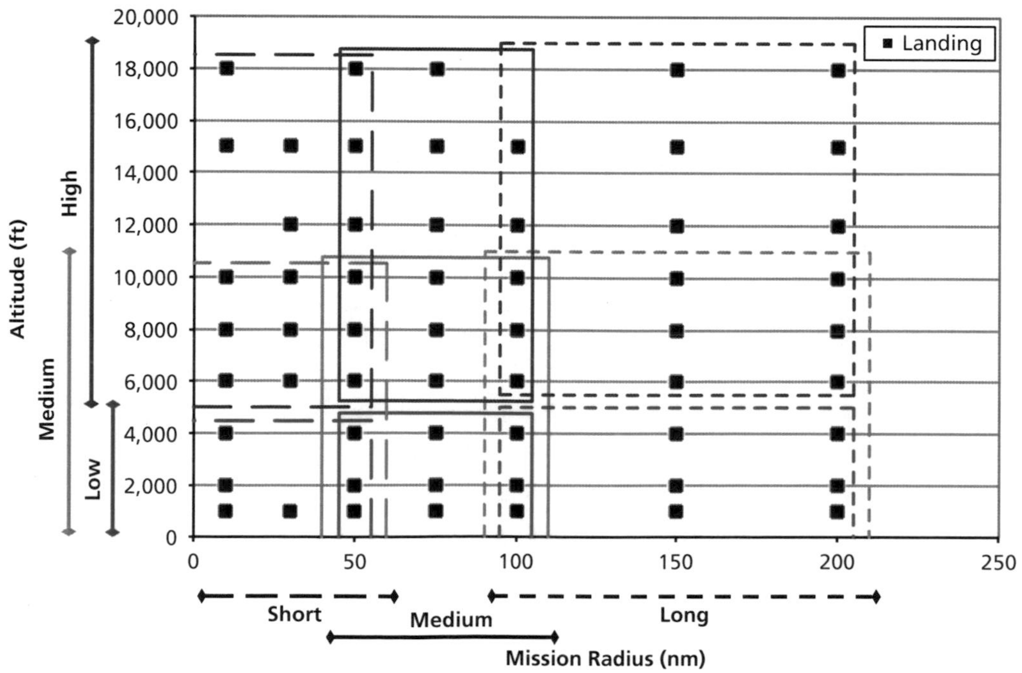

RAND *RR141z1-2.13*

CHAPTER THREE

Effectiveness Analysis

The next step in the analysis is to evaluate helicopter effectiveness. This entails establishing how many of each aircraft alternative is required to meet the demands of each of the mission-tasks defined for the 33 partner nations considered.

The Effectiveness Model

We measure helicopter effectiveness by calculating the number of aircraft needed to complete a given mission-task. To arrive at this number, we designed and applied a model with automated features capable of evaluating multiple functional and environmental parameters simultaneously. These parameters, or inputs, and the method used to arrive at a numeric output, will be described in detail.

Input Parameters

A prime determinant of the number of aircraft needed to complete a mission-task is the helicopter's payload capacity, which is defined as mission payload—one or more of equipment, supplies, passengers, litters, and so forth—and an additional 20 minutes of fuel, assuming normal cruise speed, as required by Federal Aviation Regulation §91.151.[1] The mission's delivery requirements establish the average ratio of outbound to retrograde cargo, which allows us to calculate each aircraft's maximum cargo capacity and, in turn, the number of solo-aircraft sorties required to deliver the payload in its entirety.

Payload capacity, however, is not a constant, but rather is a variable affected significantly by flying altitude, temperature, and hover performance. The model captures data for each of these. Figure 3.1 offers an illustration, for example, of how terrain feature data, derived from JOGs MEFs, translate into altitude requirements for mission-task flight paths. In this instance, the aircraft flies at 15,000 feet to clear the highest MEF of 14,300 feet, and performs a spiral climb at the LZ to reach that elevation prior to encountering the terrain obstacle. Note that the spiral climb is not explicitly depicted, but rather the climb angle is increased.

We similarly account for the effects of temperature, defining three conditions that span the range of temperatures likely to obtain in the operational environments of our country-set, up to and including extreme heat: International Standard Atmosphere (ISA), Hot, or Extra-Hot. ISA sets sea-level temperature at +15° Celsius (C); this falls at a rate of 2°C per 1,000 feet,

[1] Federal Aviation Regulation §91.151 designates fuel requirements for flight under Visual Flight Rule conditions. Federal Aviation Administration, *Code of Federal Regulations,* §91.151.

Figure 3.1
Terrain Feature Data and Mission-Task Flight Paths

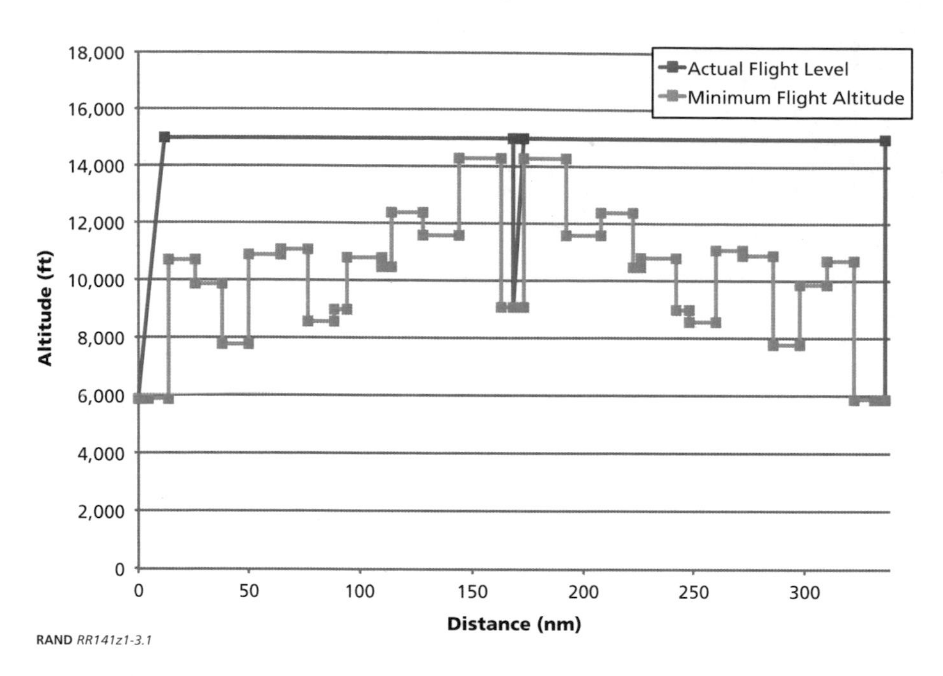

RAND *RR141z1-3.1*

up to 36,000 feet—above which temperature is assumed to take the constant value of –57°C. Hot is defined as the sea-level temperature given in DoD military standards, with a standard lapse rate; Extra-Hot is an additional 8.5°C above Hot.[2] In all cases, sea-level pressure is taken to be 101.31 kilopascals, and an ideal gas is assumed along with hydrostatic equilibrium to obtain the pressure and density profiles. Figure 3.2 shows the pressures and temperatures used for these three discrete atmospheric conditions.

We also include temperature measures with greater specificity in the form of typical daily heat. As examples, these data from August 2011 to August 2012 for Kabul, Afghanistan, and Baghdad, Iraq, are displayed in Figures 3.3 and 3.4, respectively, in relation to the ISA, Hot, and Extra-Hot thresholds described.[3] The red and blue dots indicate high and low temperatures, respectively, by day, and the red and blue lines are 15-day centered moving averages.

Finally, the effects of hover performance on payload capacity are captured through the inclusion of each mission-task's in-ground effect (IGE) and/or out-of-ground effect (OGE) hover requirements.[4]

Using these parameters as inputs, the model can identify the number of sorties needed to complete the assigned mission-task for each individual helicopter, and so compute the necessary fleet size. Recall, for example, that all mission-tasks are designed either as single-lift or as lift in one 11-hour period of daylight (daylight-lift)—from first-wheels up to last-wheels down. If the mission-task calls for single-lift delivery, then the number of sorties a single helicopter

[2] DoD Standard Practice, *Glossary of Definitions, Ground Rules, and Mission Profiles to Define Air Vehicle Performance Capability,* MIL-STD-3013, February 14, 2003.

[3] Data from Iowa Environmental Mesonet, ASOS/AWOS data download.

[4] Ground effects influence the lift and drag on the helicopter rotors. These effects exist at altitude less than about the diameter of the rotor.

Figure 3.2
Atmospheric Conditions

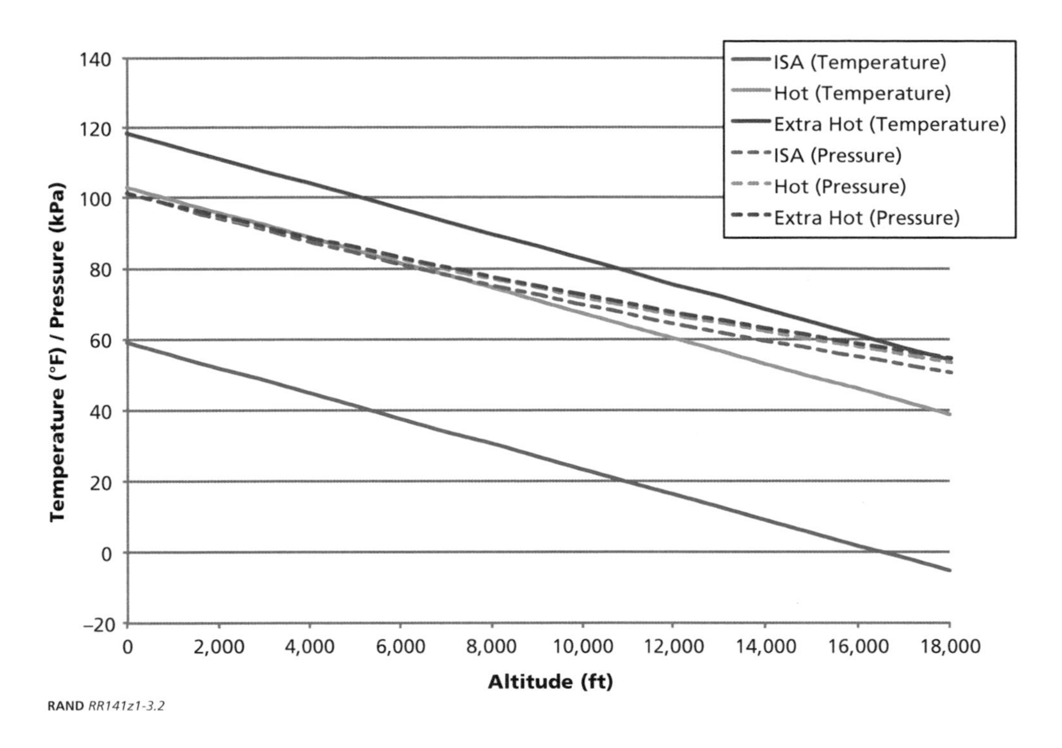

RAND *RR141z1-3.2*

flying by itself would need to deposit the payload and return to base would be equivalent to the number of aircraft needed to fulfill the mission-task. In other words, if one helicopter would need seven sorties to deliver the payload, then the single-lift mission-task would require a fleet of seven helicopters. Alternatively, if the mission-task calls for daylight-lift, we must account not only for cruise flight timc, but also for turn-time—climb time, load and unload time, and any maintenance/refueling time—to arrive at the number of sorties each aircraft alternative can complete during the designated 11-hour period.[5] The ratio of payload to sortie then allows us to calculate the total number of aircraft needed to fulfill the daylight-lift mission-task.

As described in Chapter Two, mission-tasks include direct routes and, where possible, indirect routes that avoid high-terrain features but that concomitantly increase flight distance and time. Our analysis flies all helicopter alternatives through both the direct and the alternate routes, and selects the route that maximizes effectiveness; this procedure avoids introducing bias by forcing helicopters to clear terrain features that easily can be flown around. Figure 3.5 offers one example of such alternate routing, with the red line denoting a direct route over an 11.9 MEF region and the green line a longer indirect route over lower terrain.

[5] The basic turn time at the MOB/FOB was modeled as a fixed 30 minutes. For aircraft without a cargo ramp, the load times were 20 minutes per ton, two passengers per minute, and five minutes per litter. These times were halved for aircraft with a cargo ramp. Refueling times were calculated based on fuel rates of 80 gallons per minute. For aircraft without a cargo ramp, the unload times were ten minutes per ton, four passengers per minute, and three minutes per litter; again, these times were halved for aircraft with a cargo ramp. Additional time was added for startup, taxi, and take-off (STTO): five minutes at the MOB/FOB and 2.5 minutes at the LZ.

Figure 3.3
Example Atmospheric Condition Data: Kabul

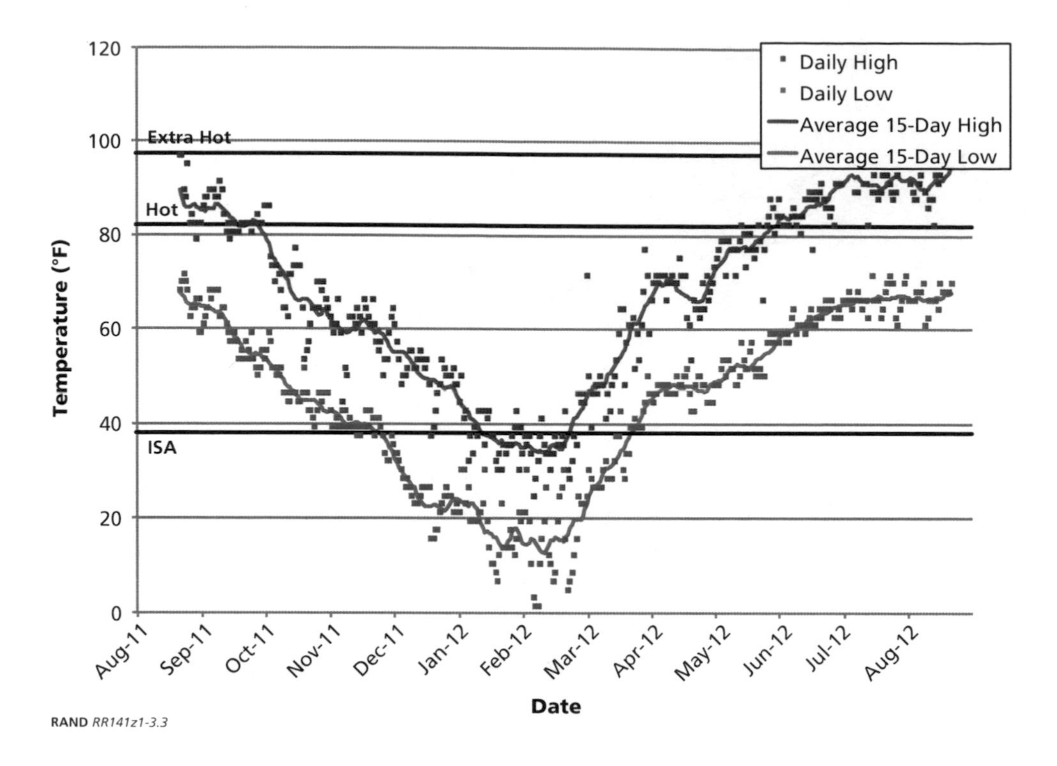

RAND *RR141z1-3.3*

Figure 3.4
Example Atmospheric Condition Data: Baghdad

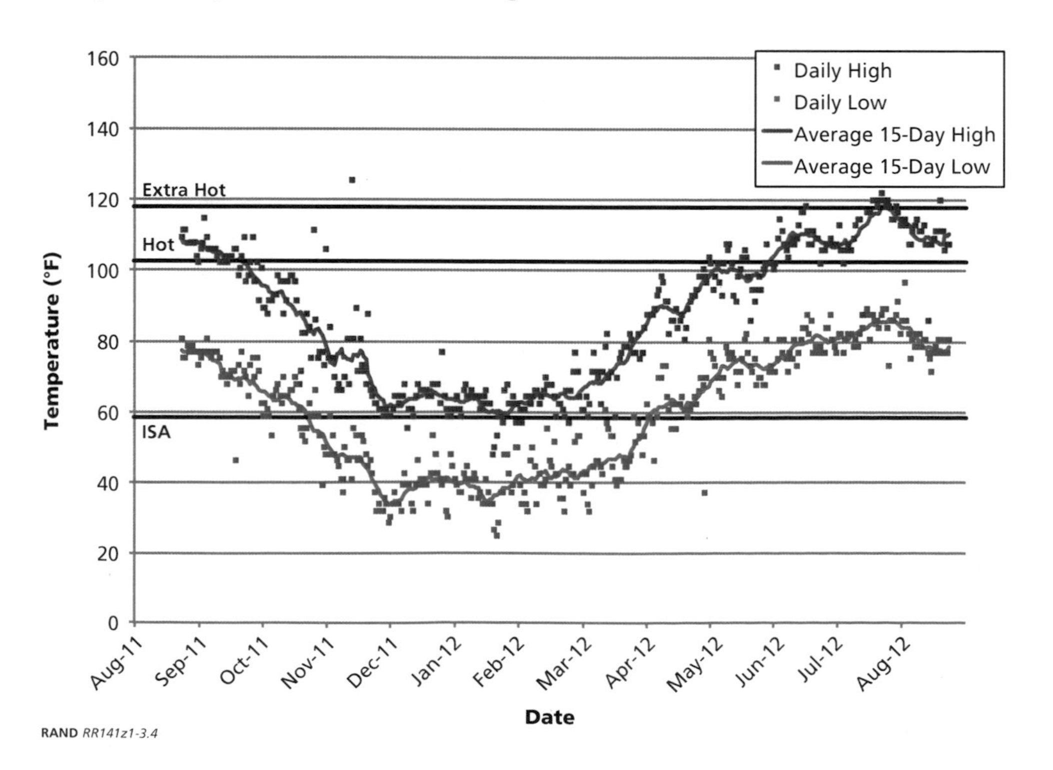

RAND *RR141z1-3.4*

Figure 3.5
Example Mission-Task Direct and Indirect Routes

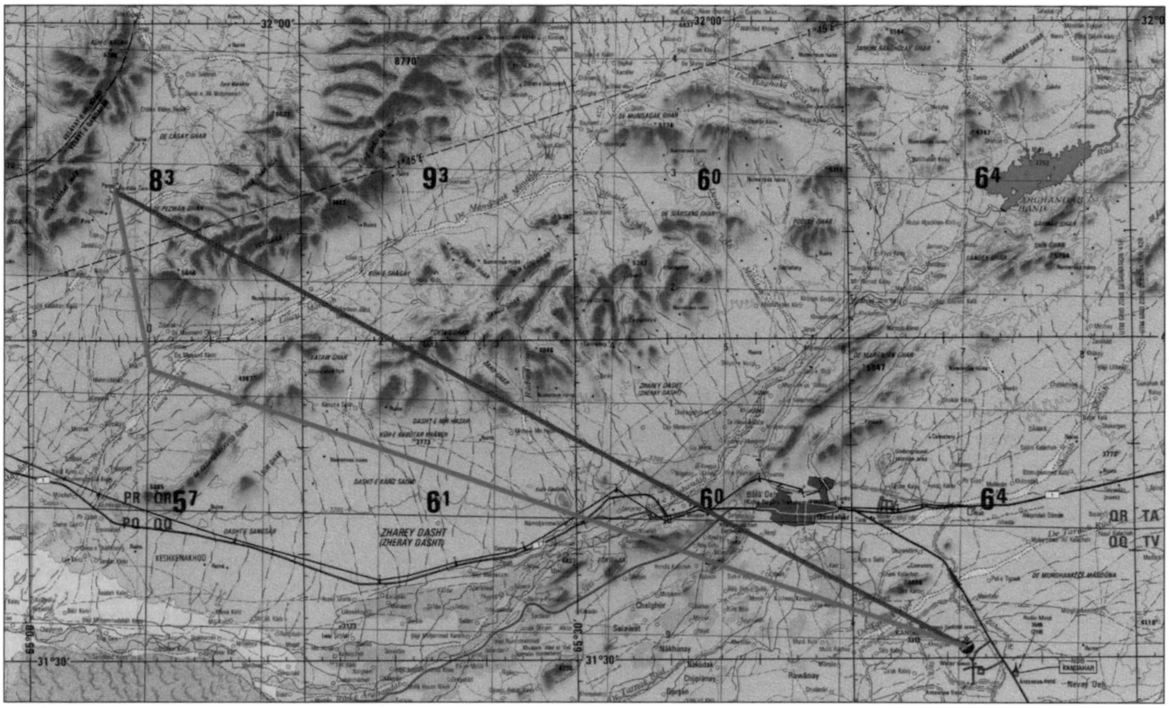

RAND *RR141z1-3.5*

Method of Calculation

As described previously, our analysis includes 27 alternative platforms and a total of 889 mobility and 651 attack mission-tasks across the 33 partner nations included in our analysis. Each helicopter must, for each mission-task, be evaluated for different combinations of mission payload, mission hover requirement, mission distance, maximum endurance speed, en route altitude, and temperature. To manage this level of complexity, we designed a model, displayed in Figure 3.6, to capture fixed mission characteristics, including weight and total elapsed time at the end of each mission segment, mission distance, and maximum altitude, as well as adjustable specifications such as the weight ratio between outbound and retrograde cargo, in-ground-effect vs. out-of-ground-effect hovers at LZs, refueling at LZs, and weight difference—indicated as "BW Plus," which is the difference between the aircraft basic weight (BW) and its operational empty weight. Flight profile is similarly adjustable, allowing selection of flight at minimum terrain-clearing altitude; maximum altitude for current weight; or altitude for maximizing range at average weight. Using these inputs, the model can calculate a helicopter's maximum payload, maximum endurance time, or maximum radius for each mission-task, and determine whether it is capable of completing the mission as specified.

In the following section, we present model outputs for lift and attack helicopters for four of the nine archetype partner nations developed in Chapter Two. These cases provide a good accounting of the relative capabilities of helicopter alternatives across a variety of mission-tasks.

Figure 3.6
Model Specifications

Mission-Task	Outbound	Inbound	Start	STTO	Climb	Cruise	Land/Endure	(Un)load	STTO	Climb	Cruise	Land
Weight (lb)	3,025	1,008	24,945	24,824	24,607	22,999	22,999	20,982	20,922	20,838	19,341	19,341
Time (hr)			0.00	0.00	0.20	1.24	1.41	1.41	0.00	1.55	2.63	2.80
Distance (nm)	168	168	0	0	12	168	168	168	168	173	336	336
Stationary Climb (nm)					0					3		
Flight Level (FL) (100-ft)	59	91	59	59	150	150	91	91	91	150	150	59
Fuel Burn (lb)				121	217	1,607	0	0	61	84	1,496	0
Distance Traveled (nm)				0	12	156	0	0	0	5	163	0
Time Traveled (hr)				0.08	0.12	1.04	0.17	0.00	0.04	0.10	1.09	0.17
Specific Range (nm / lb)					0.054	0.097				0.059	0.109	
Speed (nm / hr)					100	150				52	150	
Compute Max Payload w/ Outbound Payload Fraction	0.75		Compute Ability to Endure Hour			0.00		Compute Max Radius w/ Given Weights and at FL			150	

BW Plus	1165
M-T BATCH PROCESS	
Profile	Low
PAYLOAD RANGE CHART	
LZ Hover	OGE
Refuel	No

RAND *RR141z1-3.6*

Effectiveness Outputs, Four Archetype Partner Nations

For both the lift and the attack helicopter alternatives, we present the helicopters' overall capability under both ISA and Hot conditions, and present outputs specific to a sample mission-task. Range-payload curves for light helicopters, and range-endurance curves for attack helicopters, as well as more detailed effectiveness results, are presented in a companion document.[6]

Overall Effectiveness of Lift Helicopter Alternatives

Recall from Chapter Two that each of the nine archetype countries is representative of a "basket" of partner nations, the operational environments of which are similar in altitude and mission distance. We therefore display the model's effectiveness results not by country, but rather along these two dimensions.

Table 3.1 presents the effectiveness rating of each helicopter in each altitude-distance domain, evaluated under Hot conditions and with no fewer than four passengers per aircraft.[7] Four effectiveness ratings are coded, by color, to indicate the percentage of the set of mission-tasks for each archetype partner nation that the helicopter is capable of completing. Blue indicates completion of greater than 85 percent of the mission-tasks; green, 70–85 percent; orange, 55–70 percent; and gray, less than 55 percent.

The table reveals that although all of the helicopter alternatives are capable of executing 85 percent in the low-short domain, this effectiveness degrades markedly for some platforms as missions move into high altitudes and/or long ranges. Indeed, where mission-tasks are both of high altitude and of long range, only four of the 24 lift helicopter alternatives achieve 85 percent effectiveness, with an additional two capable of completing at least 70 percent of the partner nation's mission-tasks.

Sample Mission-Task Effectiveness of Lift Helicopter Alternatives

Because these analyses are standardized—with the same set of mission-tasks run under the same conditions for each of the archetype partner nations—the model allows us to acquire a

[6] Mouton et al., 2014, not available to the general public.

[7] Results for other temperature conditions are presented in Mouton et al., 2014, not available to the general public.

Table 3.1
Lift Helicopter Effectiveness for Archetype Partner Nations Under Hot Conditions

	Short				Long			
High	Bell 407	Bell 429	Huey II	UH-1N	Bell 407	Bell 429	Huey II	UH-1N
	Bell 412EP	UH-1Y	CH-46E	CH-47D	Bell 412EP	UH-1Y	CH-46E	CH-47D
	CH-47 Int	S-61T	S-61T+	UH-60L	CH-47 Int	S-61T	S-61T+	UH-60L
	UH-60M	S-92	AW109	AW139	UH-60M	S-92	AW109	AW139
	AW149	EH101	EC-145	AS-332L1	AW149	EH101	EC-145	AS-332L1
	NH-90	EC-725	Mi-17	Mi-17v5	NH-90	EC-725	Mi-17	Mi-17v5
Low	Bell 407	Bell 429	Huey II	UH-1N	Bell 407	Bell 429	Huey II	UH-1N
	Bell 412EP	UH-1Y	CH-46E	CH-47D	Bell 412EP	UH-1Y	CH-46E	CH-47D
	CH-47 Int	S-61T	S-61T+	UH-60L	CH-47 Int	S-61T	S-61T+	UH-60L
	UH-60M	S-92	AW109	AW139	UH-60M	S-92	AW109	AW139
	AW149	EH101	EC-145	AS-332L1	AW149	EH101	EC-145	AS-332L1
	NH-90	EC-725	Mi-17	Mi-17v5	NH-90	EC-725	Mi-17	Mi-17v5

Legend

Percentage of Missions Completed with Four or More Passengers per Aircraft			
≤55%	55% – 70%	70% – 85%	≥85%

Altitude Bands (ft)

	Low	High
MOB	1,000 – 4,000	7,000 – 13,000
En Route	1,100 – 6,000	6,600 – 21,600
LZ	1,000 – 4,000	6,000 –

Distance Bands (nm)

	Short	Long
Distance	10 – 50	100 – 200

general measure of overall helicopter capability, and to identify the factor for each platform that most often limits its payload capacity. Four such factors are fairly common across the helicopter alternatives: maximum gross takeoff weight (MGTOW); fuel carrying capacity; inability to IGE hover at all or, for sufficient duration, at the initiating base; and the inability to OGE hover at the LZ. The most common of these is an inability to OGE hover at the LZ. Table 3.2 associates each helicopter alternative with its most significant limiting factor. MGTOW is indicated in purple; fuel carrying capacity in gold; IGE hover at the MOB in blue; and OGE hover at the LZ in magenta.

Overall Effectiveness of Attack Helicopter Alternatives

We subjected the attack helicopter alternatives to a similar set of analyses, beginning again with general performance in completing mission-tasks for each of the altitude-distance "baskets" under Hot conditions, with no fewer than four passengers per aircraft. For attack platforms, we add the requirement of a minimum 15-minute loiter; performance measures are coded as follows: blue indicates an ability to meet the 15-minute loiter threshold in at least 85 percent of the missions in the mission-task set; green, 70–85 percent; orange, 55–70 percent; and gray, less than 55 percent. Aircraft results are presented in Table 3.3.

In the low-short domain, all aircraft are able to loiter for at least 15 minutes in at least 85 percent of the mission-set. Missions that require high altitudes and/or long ranges produce capability degradation in three of the four alternatives considered; only the AH-1Z achieves greater than 85 percent capability in all distance-altitude domains.

Table 3.2
Lift Helicopter Capability in Archetype Partner Nations Under Hot Conditions

High, Short:

Bell 407	Bell 429	Huey II	UH-1N
Bell 412EP	UH-1Y	CH-46E	CH-47D
CH-47 Int	S-61T	S-61T+	UH-60L
UH-60M	S-92	AW109	AW139
AW149	EH101	EC-145	AS-332L1
NH-90	EC-725	Mi-17	Mi-17v5

High, Long:

Bell 407	Bell 429	Huey II	UH-1N
Bell 412EP	UH-1Y	CH-46E	CH-47D
CH-47 Int	S-61T	S-61T+	UH-60L
UH-60M	S-92	AW109	AW139
AW149	EH101	EC-145	AS-332L1
NH-90	EC-725	Mi-17	Mi-17v5

Low, Short:

Bell 407	Bell 429	Huey II	UH-1N
Bell 412EP	UH-1Y	CH-46E	CH-47D
CH-47 Int	S-61T	S-61T+	UH-60L
UH-60M	S-92	AW109	AW139
AW149	EH101	EC-145	AS-332L1
NH-90	EC-725	Mi-17	Mi-17v5

Low, Long:

Bell 407	Bell 429	Huey II	UH-1N
Bell 412EP	UH-1Y	CH-46E	CH-47D
CH-47 Int	S-61T	S-61T+	UH-60L
UH-60M	S-92	AW109	AW139
AW149	EH101	EC-145	AS-332L1
NH-90	EC-725	Mi-17	Mi-17v5

Legend

Percentage of Missions Completed with Four or More Passengers per Aircraft			
MGTOW	Fuel Limit	MOB Hover	LZ Hover

Altitude Bands (ft)

	Low	High
MOB	1,000 – 4,000	7,000 – 13,000
En Route	1,100 – 6,000	6,600 – 21,600
LZ	1,000 – 4,000	6,000 – 18,000

Distance Bands (nm)

	Short	Long
Distance	10 – 50	100 – 200

Table 3.3
Attack Aircraft Effectiveness for Archetype Partner Nations Under Hot Conditions

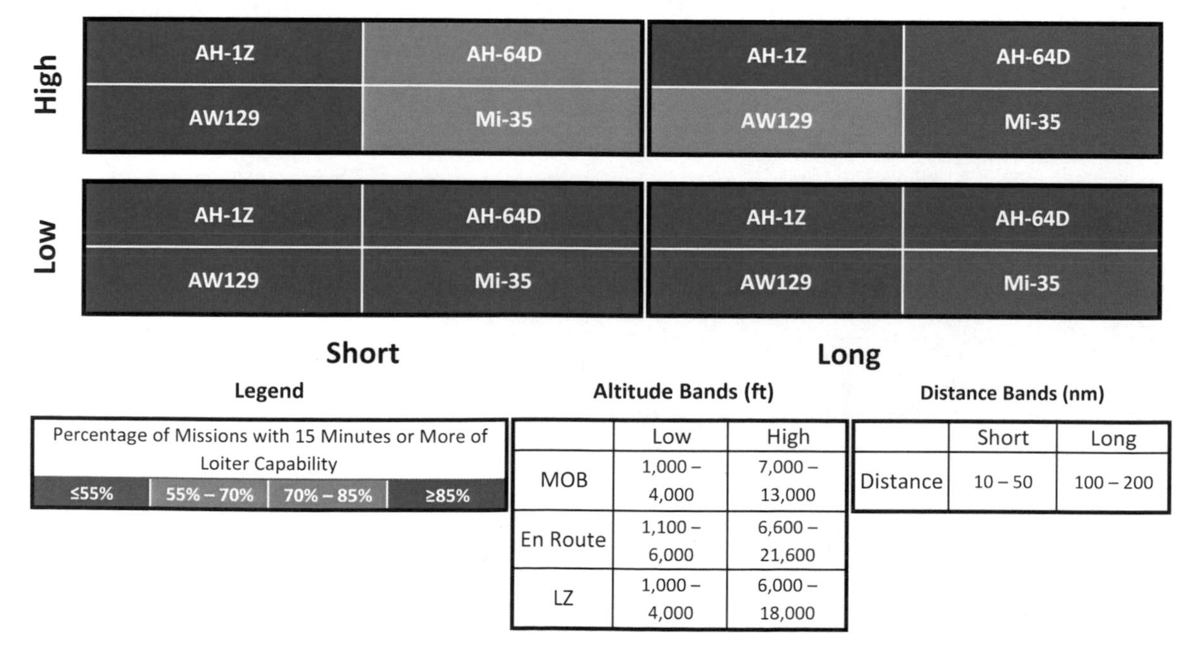

High, Short:

AH-1Z	AH-64D
AW129	Mi-35

High, Long:

AH-1Z	AH-64D
AW129	Mi-35

Low, Short:

AH-1Z	AH-64D
AW129	Mi-35

Low, Long:

AH-1Z	AH-64D
AW129	Mi-35

Legend

Percentage of Missions with 15 Minutes or More of Loiter Capability			
≤55%	55% – 70%	70% – 85%	≥85%

Altitude Bands (ft)

	Low	High
MOB	1,000 – 4,000	7,000 – 13,000
En Route	1,100 – 6,000	6,600 – 21,600
LZ	1,000 – 4,000	6,000 – 18,000

Distance Bands (nm)

	Short	Long
Distance	10 – 50	100 – 200

CHAPTER FOUR

Cost Analysis

In this chapter, we present the methodology used to generate the cost for each aircraft alternative fleet analyzed. Standard cost-estimating procedures were used in this analysis. Throughout this chapter, we provide a considerable amount of detail on the approach and underlying data.

First, we present the background on the overall cost approach, which includes the key costing assumptions. The total life-cycle cost of an aircraft depends on both the flyaway cost of the aircraft and its O&S cost. Details on the costing approach for both of these are presented. In addition, we discuss the maintainability of Russian versus Western helicopters.

Definitions

Because the aircraft analyzed are existing designs currently in production, we have excluded from our analysis the costs of development that usually would be included in an aircraft's "cradle-to-grave" life cycle. Instead, we define the total acquisition cost of a helicopter fleet as being the summation of the costs incurred during the processes of production, and the initial set of support materials and training programs needed to bring the aircraft into service. Production refers to the making of the helicopter, and specifically to the funds needed to acquire and to use the tools, and to complete the human and mechanical processes, needed to manufacture each aircraft. We refer to these as "flyaway costs," and add to them "initial O&S costs" (the price of the support items needed to initiate operations and maintenance activities)—e.g., a first set of spare and repair parts, maintenance equipment specific to the aircraft design, manuals and educational materials, the training of an initial class of pilots and maintenance staff, and field activities. We rely upon the May 1, 1992, OSD-CAIG structure to define the categories of O&S to be included in our model.[1] These support items together with flyaway are defined in our model as the total acquisition cost of an aircraft.

We acquire the flyaway costs for each platform from a variety of sources, described in full later, and add to them initial O&S costs at a standard 12 percent, a figure that represents the average value of the ratio of procurement to flyaway cost.[2] All costs are expressed in constant

---

[1] Office of the Secretary of Defense Cost Analysis Improvement Group, 1992. This structure is consistent with that of the AFTOC.

[2] This value is the average of the ratios of procurement cost to flyaway cost of Army helicopter programs using data from the December 2009 Selected Acquisition Reports (SARs) and the 2011 President's Budget exhibits. These include the AH-64, CH-47, UH-60, and UH-72. DoD, Defense Acquisition Management Information Retrieval (DAMIR), *Selected Acquisition Report (SAR): RCS: DD-A&T(Q&A)823-182 (LUH),* December 31, 2009a; DoD, DAMIR, *Selected Acquisition Report, RCS: DD-A&T(Q&A)823-278 (CH-47F),* December 31, 2009b; DoD, DAMIR, *Selected Acquisition Report (SAR):*

fiscal year (FY) 2010 dollars, either having been obtained as such or having been calculated using the appropriate military service escalation tables from Air Force raw index values for aircraft procurement.[3] Where needed, current-value calculations were made using a 2.7 percent real discount rate, and a 20-year cost horizon.[4]

All aircraft included in the analysis are modeled as flying 353 hours per year, which is equivalent to 300 flying hours per total inventory aircraft. Personnel for all helicopters was set at 1.2 crews per aircraft, a number that would be low by U.S. standards but that SMEs report is an appropriate figure for partner nations. Ten percent of the operating fleet is devoted to training, and mission-capable rate is set at 75 percent, which is the U.S. Air Force and U.S. Army average across helicopter fleets. The depot-possessed rate is consistent with the U.S. average at 15 percent.

Flyaway Cost

We obtained flyaway cost data from various outlets, including the USG, contractors, and trade press and other open-source media. For the Mi-17v5, we relied upon recent contracts and databases maintained by the USG, as well as two USG reports.[5] We requested price information from all of the contractors producing the helicopter alternatives included in our analysis, and received responses from Bell, Boeing, Sikorsky, and Eurocopter. Where available, we accepted the valuation of the helicopters' commercial variants as provided by aviation consultancy Conklin & de Decker. In cases where even this information was unavailable, we used a "comparable data" approach (labeled "CER," for cost-estimating relationship); we were unable to arrive through any of these methods at a reliable estimate for three helicopters: the AW129, the Mi-35, and the Mi-17 (HIP-H). For all aircraft for which they were available, USG data were included in the model unless the contractor-provided estimate was higher.[6]

Bell Flyaway Cost-Estimates

Table 4.1 provides the flyaway cost values for the Bell helicopters that were used in our analysis. Source is indicated by column, with the final input values highlighted in blue.

Bell provided values for four of the helicopters: the Bell 407, the Bell 429, the Huey II, and the Bell 412EP. Bell did not provide the price of the UH-1N, but because the DoS report included values both for this platform and for the Huey II, we were able to use the ratio

RCS: DD-A&T(Q&A)823-341 (UH-60M), December 31, 2009c; DoD, DAMIR, *Selected Acquisition Report (SAR): RCS: DD-A&T(Q&A)823-831 (Longbow Apache),* December 31, 2009d; Department of the Army, Procurement Programs, *Fiscal Year 2011 Budget Estimate: Aircraft Procurement,* February 2010.

[3] U.S. Air Force, *AFI 65-503, Cost Factors,* web page, undated. Not available to the general public.

[4] This discount value is from December 2009, and valid for FY 2010. See Office of Management and Budget, *Discount Rates for Cost-Effectiveness, Lease Purchase, and Related Analyses,* OMB Circular No. A-94, Appendix C, December 2011. The OMB Circular presents nominal and real rates. The nominal rates are used with then-year (includes the effects of inflation) dollars; the real rates are used with constant-dollar values. The values vary with the time horizon of the analysis.

[5] DoD, *Medium-Lift Helicopter Requirements for the Afghanistan National Army Air Corps (ANAAC),* Operational Requirements and Analysis Document, February 11, 2010a, p. 13; DoS, 2009.

[6] We saw no evidence to suggest contractor prices were biased, either up or down. Details on our flyaway cost estimates are presented in Mouton et al., 2014, not available to the general public.

Table 4.1
Sources and Platform Cost-Estimates ($M FY 2010): Bell

Platform	MGTOW (lb)	Contractor[a]	USG[b]	Conklin & de Decker	DoD Analysis of Alternatives (AoA)[c]	DoS	CER[a]	Other[d]
AH-1Z	18,500		30.9		29.0			
Bell 407	5,250	—		2.5				
Bell 429	7,000	—						
Huey II	10,500	—				3.6		7.0
UH-1N	10,500			3.6		3.8	—	
Bell 412EP	11,900	—				8.7		
UH-1Y	18,500		21.2		21.0			

[a] Information provided in Mouton et al., 2014, not available to the general public.

[b] Department of the Army, 2010, Exhibit P-5.

[c] DoD, 2010b.

[d] TH-1H cost, received from the Air Force Acquisition Office, April 16, 2010.

between the Bell Huey II and the DoS Huey II estimates to scale similarly the cost of the UH-1N. We take USG data for the remaining two Bell models, the AH-1Z and the UH-1Y.

Boeing Flyaway Cost-Estimates

Table 4.2 provides the flyaway cost values for the Boeing helicopters that were used in our analysis. Our estimate for the AH-64D is based on USG budget documents. For the CH-47D and CH-47 International, we used contractor data and a RAND analysis of alternatives to calculate what we believe to be accurate estimates. For the CH-46E, we did a regression of cost on empty weight for the Bell 407, AW109, Bell 429, EC-145, Bell 412EP, AW139, UH-1Y, UH-60L, UH-60M, NH90, Super Cougar (EC-725), S-92, EH101, Huey II, UH-1N, S-61T, S-61T+, AS-332L1, and CH-47D. We included a dummy for a refurbished model, which we applied to the last six helicopters. Specifically, the regression equation was of the form

$$\ln(\text{cost})=\alpha+\beta \ln(EW)+\gamma\delta_{refurb},$$

where α, β, and γ are the regression coefficients, cost is the platform cost in millions of dollars, EW is the aircraft empty weight in pounds, and δ_{refurb} is 1 if the aircraft is refurbished and 0 otherwise. In this regression, the constant coefficient, α, is –7.77 (t-statistic = –6.01); the coefficient on weight, β, is 1.14 (t-statistic = 7.98); the coefficient on the refurbishment dummy, γ, is –0.39 (t-statistic = –2.25); R-square = 0.80; standard error = 0.35. The cost estimate

Table 4.2
Sources and Platform Cost-Estimates ($M FY 2010): Boeing

Platform	MGTOW (lb)	Contractor	USG	Conklin & de Decker	DoD AoA	DoS	CER	Other
AH-64D	20,260		32.2[a]		32.0			
CH-46E	24,300							16.6[b]
CH-47D	50,000	_[c, d]						
CH-47 Int	54,000							_[d]

[a] Department of the Army, *Fiscal 2010 Budget Estimate, Aircraft Procurement,* 2009, Exhibit P-3A.

[b] This is the CER estimate of a what a completely refurbished CH-46E would cost. We also analyze its cost-effectiveness at a minimal acquisition cost.

[c] A CH-47F, a remanufactured CH-47D with a 20-year life, has U.S. government-reported cost of $22.4 million (Department of the Army, *Fiscal 2012 Budget Estimate, Aircraft Procurement*, 2011, Exhibit P-5) and $23.7 million (DoD, DAMIR, 2009b).

[d] Information provided in Mouton et al., 2014, not available to the general public.

included in the model, $16.6 million, is for a completely refurbished CH-46E; as a sensitivity check we also assess cost-effectiveness at a minimal acquisition cost—wherein the helicopter is essentially given to a partner nation without any refurbishment. Whether such aircraft would have 20 years of life is uncertain, and so this check errs on the side of overestimating the CH-46E's cost-effectiveness. Source is indicated by column, with the final input values highlighted in blue.

Sikorsky Flyaway Cost-Estimates

We used Sikorsky-provided cost-estimates for all of its helicopters with the exception of the UH-60M. For that model, we used the value indicated in the President's Budget rather than that indicated in the SAR;[7] we did so because the former is a forward-looking document rather than a backward-reporting one, and so better captures future expenditures. Table 4.3 provides these flyaway cost-estimates, with source again indicated by column and final input values highlighted in blue.

AgustaWestland Flyaway Cost-Estimates

AgustaWestland did not contribute cost-estimates for its helicopters. We therefore relied heavily on data provided by Conklin & de Decker. We estimated cost for the AW149 by scaling to the cost of the AW139, using the ratio of their empty weights taken to the weight-coefficient power calculated for the Boeing CH-46E: AW139 Empty-Weight$^{(\beta=1.14)}$/AW149 Empty-Weight$^{(\beta=1.14)}$. Although we did find price estimates on a trade press website, we were unable to verify their credibility; because this was the only source of data on the AW129, we opted instead to peg the cost at which an AW129 would be competitive with American-

[7] DoD, DIMR, 2009c; Department of the Army, 2010.

Table 4.3
Sources and Platform Cost-Estimates ($M FY 2010): Sikorsky

Platform	MGTOW (lb)	Contractor[a]	USG	Conklin & de Decker	DoD AoA	DoS	CER	Other
S-61T	20,500	—			9.0[b]			13.6[c]
S-61T+	20,500	—						
UH-60L	22,000	—			10.0[d]			
UH-60M	22,000		14.9[e] 16.4[f]		15.0			
S-92A	26,500	—		23.7		19.5		

[a] Information provided in Mouton et al., 2014, not available to the general public.

[b] DoD AoA refers to the S-61N in general, and says this price is for an aircraft that is "not military." DoD, 2010b, p. 43.

[c] Based on average contract cost ($15.2M) of DoS S-61T (with Carson composite rotor blade) buy. This may include some additional logistic support. Giovanni de Briganti, "U.S. State Dept. Order for 110 S-61Ts Is Boon for Sikorsky," *Rotor and Wing,* May 1, 2010, UNCLASSIFIED.

[d] This is for refurbishment of a UH-60A. There are no UH-60As available for this in the time frame of this study, and the UH-60L for this study is a new S-70i manufactured in Poland.

[e] DoD, DAMIR, 2009c.

[f] Department of the Army, 2010, Table P-5.

made attack helicopters, as derived from performance in completing mission-tasks (see Chapter Two). To arrive at this figure, we apply the CER described earlier to the Bell AH-1Z and the Boeing AH-64D. The result is under-prediction by 59 percent—that is, we put the regression costs for these helicopters at 59 percent less than the price recorded by the USG. This factor is then used to derive the "CER" estimate for the AW129. Table 4.4 provides all collected flyaway cost-estimates, sources, and final input values.

Eurocopter Flyaway Cost-Estimates

We used contractor inputs for three of the Eurocopter helicopters, and the Conklin & de Decker data source for the AS-332L1. These are displayed in Table 4.5.

Mil Moscow Helicopter Plant Flyaway Cost-Estimates

Our Mi-17v5 value is derived from both a recent contract for new Mi-17v5s and a database from the field on recent purchases. Application of the CER produces a cost overestimation of 53 percent; this suggests that the Mi-17v5 is less expensive, based on weight, than its Western counterparts. We found no reliable estimate for the cost of the Mi-35. This, as well as the data collected for the Mi-17v5, is displayed in Table 4.6.

Table 4.4
Sources and Platform Cost-Estimates ($M FY 2010): AgustaWestland

Platform	MGTOW (lb)	Contractor	USG	Conklin & de Decker	DoD AoA	DoS	CER	Other
AW109	6,614			6.7				9.0[a]
AW129	11,660						17.1	63.0[b]
AW139	13,232			12.6		12.4		10.0[c]
AW149	16,000						14.3[d]	13.0[e]
EH101	34,392			21.0		20.6		25.0[f]

[a] Aircraftcompare.com, *Agusta-Westland—United Kingdom, Military AW109 LUH*, undated-a.

[b] Aircraftcompare.com, *Military AW129 Mangusta*, undated-b.

[c] Aircraftcompare.com, *AW139*, undated-c.

[d] Scaled from AW139 using CER weight coefficient.

[e] Aircraftcompare.com, *AW149*, undated-d.

[f] Aircraftcompare.com, *AW101 Merlin*, undated-e.

Table 4.5
Sources and Platform Cost-Estimates ($M FY 2010): Eurocopter

Platform	MGTOW (lb)	Contractor[a]	USG	Conklin & de Decker	DoD AoA	DoS	CER	Other
EC-145	7,904	—	5.0[b] 5.6[c]	7.8				
AS-332L1	20,613			17.0		14.1		22.2[d]
NH-90	23,369	—						4.30[e]
EC-725	24,691	—						

[a] Information provided in Mouton et al., 2014, not available to the general public.

[b] DoD, DAMIR, 2009a.

[c] Department of the Army, 2011, Exhibit P-5.

[d] Jane's All the World Aircraft, *Eurocopter EC 225 and EC 725*, web page, November 20, 2009.

[e] NSRW Working Group deliberations, SOCOM Global Synchronization Conference, MacDill Air Force Base, Fla., 2010.

Table 4.6
Sources and Platform Cost-Estimates ($M FY 2010): Mil Moscow

Platform	MGTOW (lb)	Contractor	USG	Conklin & de Decker	DoD AoA	DoS	CER	Other
Mi-35	25,353						17.2	22.0[a] 16.7[b]
Mi-17v5	28,660		13.9[c]		8.0-16.0	12.4		

[a] Rough estimate based on press report of $30 million per aircraft for Brazil buy, which includes some continued maintenance. Flight International, "Brazilian Air Force Fields First Mi-35M Attack Helicopters," Flightglobal website, April 22, 2010.

[b] System Program Office says "slightly more than Mi-17V5;" this is 20 percent more. Weight-scaling from the Mi-17v5 gives a 17.5-percent increase, or a price of $16.3 million. NSRW, 2010.

[c] ARINC contract for nine Mi-17v5. ARINC, Mi-17v5 contract, W9113M-07-D-0009, October 23, 2008. This is consistent with the latest cost data provided by the NSRW System Program Office. Average procurement unit cost is $15.7 million.

O&S Costs

While there largely is consistency across sources providing data on procurement price, the same cannot be said for O&S costs. There is great variation in how dollar-per-flying-hour O&S estimates are calculated: Some include maintenance labor where others do not; other such variables for inclusion are sustaining engineering support, indirect costs such as personnel training, and base support costs (such as guards and infrastructure maintenance). As a result, it is not possible to meaningfully evaluate O&S estimates across sources for consistency. Although some data repositories (for example, Conklin & de Decker) do provide consistent estimates of O&S costs for a number of platforms, their accounting does not include the full set of helicopter alternatives being analyzed here.

We therefore use the May 1, 1992, OSD-CAIG structure to define the categories of O&S to be included in our model.[8] Although the OSD-CAIG provides the same seven top-level cost elements, the U.S. Air Force, Army, and Navy/Marine Corps each has implemented these guidelines differently.[9] The Army does not report data for all categories, the Air Force includes some number of lower-level costs the Navy excludes, and vice versa; moreover, Navy entries reflect a mix of ship-borne and land-based operations while the Air Force deals only with the

---

[8] This structure is consistent with that of the Air Force Total Ownership Cost (AFTOC).

[9] 1.0 Mission Personnel; 2.0 Unit-Level Consumption; 3.0 Intermediate Maintenance; 4.0 Depot Maintenance; 5.0 Contractor Support; 6.0 Sustaining Support; 7.0 Indirect Support.

latter. We take as the basis for our model the AFTOC structure, supplementing with Army and Navy/Marine Corps helicopter maintenance data when available.[10]

This structure is designed to capture the characteristics of the helicopters in use by the U.S. military; as such, we did not assume it would be equally applicable to the Mi-17v5 and the Mi-35, which are the Mil Moscow Helicopters included in the RAND analysis. In the following section, therefore, we describe the method by which O&S cost-estimates were made for these aircraft.

O&S Cost-Estimates for Mil Moscow Helicopter Plant Platforms

To address the potential incompatibility of the AFTOC structure and Mil Moscow Aircraft, we began by seeking to characterize the relative complexity of the aircraft. Drawing on training course data from DoD and DoS, we found that the Advance Crew Qualification course for the Mi-17 lasts six weeks, compared to four weeks for the UH-1 Huey, eight for the UH-60 Blackhawk, and ten for the CH-47 Chinook.[11] Basic maintenance technician courses last three weeks for the Mi-17, four for the UH-1, six for the UH-60, and eight for the CH-47.[12] These data suggest that the Mi-17 is broadly less complex to fly and maintain than contemporary U.S. helicopters.

We then solicited direct impressions and data from SMEs, most particularly from those U.S. military personnel with direct and regular experience in Afghanistan and Iraq. Our discussions with SMEs indicate that the Mil Moscow systems being analyzed here have no pronounced advantage or disadvantage in overall maintainability and reliability relative to contemporary Western models. Unit-level maintenance of Russian helicopters tends to involve more, but much simpler, maintenance actions over a given period of time than contemporary Western helicopters. This suggests that perhaps Russian helicopters are easier for a less-educated workforce to maintain. However, they also require much more frequent depot-level overhauls than Western helicopters, which present a host of issues. While the maintenance construct is very different, the overall maintainability and reliability is not, as reflected in the fact that the mission-capable rate for Mi-17s in the U.S. units we visited was comparable to Western aircraft maintained by those and other units.

This result is consistent with recent DoD and DoS reports, which provide similar estimates for O&S costs for the Mi-17v5 and its Western alternatives. The compatibility of these findings led us to conclude that the AFTOC structure could usefully be applied to these aircraft.

O&S Cost Model

We have divided the cost elements in the OSD-CAIG structure into two parts. The first we call "direct costs"—those that are closely linked to operational activity. These include crew personnel, maintenance personnel, fuel, consumables (maintenance materials), depot-level reparables (equipment items), training expendables (primarily munitions), and depot maintenance (airframe and engine overhauls). The Air Force does not report costs for intermediate maintenance, but the Navy and Army do; we subsume these in the maintenance personnel category.

---

[10] Army Operating and Support Management Information System (OSMIS) website, undated. Navy Visibility and Management of Operation and Support Cost (VAMOSC) website, undated. These websites are not available to the general public.

[11] DoS and DoD, *Foreign Military Training and DoD Engagement Activities of Interest 2011-2012,* Joint Report to Congress, 2012; U.S. Air Force, *Mi-17 Program of Instruction,* USAF Air Education and Training Command, 2012.

[12] DoS and DoD, 2012; U.S. Air Force, 2012; Boeing, CH-47 maintainer course program of instruction, undated.

We refer to the second component of the OSD-CAIG structure as "indirect costs." These are the costs associated with the service and command organizations whose activities support operations: staff and security, services and transportation costs attributable to the unit operations, sustaining engineering and software support, and personnel and facility-related costs attributable to unit operations.

To arrive at estimates for each of the direct- and indirect-cost OSD-CAIG categories, we use a number of sources and methods. For two of the direct-cost categories, we use outputs from the effectiveness analysis conducted in Chapter Three: crew costs and fuel costs. Crew costs are calculated using the identified 1.2 crews per aircraft, accounting for officers and enlisted at AFI 65-503 pay rates. Fuel costs reflect consumption in gallons per hour, as derived from the effectiveness analysis, at a cost (FY 2010$) of $2.54 per gallon; this number appears in the recent U.S. Air Force KC-X (aerial refueling tanker) Request for Proposals.[13]

For the other four direct-cost categories, we use CERs developed using historical Army, Navy, and Air Force helicopter O&S data from AFTOC, the Navy's Visibility and Management of Operating and Support Costs system, and the Army's Operating and Support Management Information System.[14]

Indirect-cost elements are estimated as ratios to related direct-cost elements. We estimate them by their average ratio to the associated direct-cost elements in all AFTOC data; this is consistent with our choice of the specific AFTOC implementation of the OSD-CAIG cost-element structure as our O&S cost framework. Table 4.7 displays each OSD-CAIG cost category and associated AFTOC number; the historical (FY 1996–2009) average percentage cost

Table 4.7
O&S Cost Categories and Method of Estimation

Cost Category (AFTOC #)	AFTOC Share	Estimation Method
Crew Personnel (1.1.1 + 1.1.2 + 1.1.3)	15	Effectiveness Analysis
Command and Control Personnel (1.1.4)	2	Ratio to Crew based on AFTOC
Maintenance Personnel (1.2)	23	CER
Other Mission Personnel (1.3)	5	Ratio to Other Personnel based on AFTOC
Fuel (2.1)	3	Effectiveness Analysis
Consumables (2.2)	9	CER
Depot Level Repairs (DLR) (2.3)	14	CER
Training Expendables (2.4)	0.3	AFTOC Value
Other Unit Level Consumption (2.5)	9	Ratio to Total Personnel based on AFTOC
Depot Maintenance (not DLRs) (4.0)	6	CER
Contractor Support (5.0)	2	Ratio to Other Maintenance based on AFTOC
Sustaining Support (6.0)	4	Ratio to Other Maintenance based on AFTOC
Indirect Support (7.0)	8	Ratio to Total Personnel based on AFTOC

[13] FedBizOpps.gov, *KC-X Tanker Modernization Program Request for Proposals (Solicitation FA8625-10-R-6600-SpecialNotice),* web page, September 25, 2009.

[14] The Army's OSMIS data were used for only for the consumables and depot-level-reparable CERs.

of each cost category across all U.S. Air Force helicopters (AFTOC Share); and the method of estimation.

CERs for O&S Maintenance Cost Categories

Our CER model relies upon inputs from 20 helicopters, attack and utility. These are identified in Table 4.8.

We estimated four models with different dependent variables. The first model estimated for effects on maintenance personnel per fleet size, or total active inventory. The other three cost category dependent variables—consumables, depot-level reparables, and depot maintenance (not depot-level reparables)—were measured per flying-hour. In all four models, we estimated cost as a function of aircraft empty weight, a proxy for its size, with an attack helicopter dummy variable. A logarithmic functional form was used, as is standard practice in cost estimation. The statistical results for the CER models are provided in Table 4.9.

Table 4.8
Helicopters Used in O&S Maintenance CERs

Platform	Service	Source	Flying Hours/ Total Active Inventory (TAI)/Year
AH-1W	Navy	VAMOSC	224
CH-46D	Navy	VAMOSC	428
CH-46E	Navy	VAMOSC	241
CH-53D	Navy	VAMOSC	197
CH-53E	Navy	VAMOSC	198
UH-1N	Navy	VAMOSC	221
UH-1Y	Navy	VAMOSC	160
UH-3H	Navy	VAMOSC	304
UH-46D	Navy	VAMOSC	461
AH-1F	Army	OSMIS	60
AH-1S	Army	OSMIS	33
AH-64A	Army	OSMIS	160
AH-64D	Army	OSMIS	293
CH-47D	Army	OSMIS	170
CH-47F	Army	OSMIS	294
UH-1H	Army	OSMIS	99
UH-60L	Army	OSMIS	237
UH-60M	Army	OSMIS	268
UH-1H	USAF	AFTOC	221
UH-1N	USAF	AFTOC	303

Table 4.9
CER Results for O&S Categories

CER	Weight Coefficient	Attack Dummy	Attack Multiplier	Standard Error	R-Square
Maintenance Personnel	0.44 (3.5)	0.02 (0.1)	1.02	0.22	0.60
Consumables	0.62 (2.4)	0.25 (0.8)	1.28	0.60	0.25
Depot Level Repairs (DLR)	1.11 (5.1)	0.40 (1.6)	1.50	0.46	0.64
Depot Maintenance (not DLRs)	0.92 (3.3)	–0.20 (0.4)	0.82	0.50	0.61

NOTE: t-statistic shown in parentheses.

Weight coefficients all fall in the expected magnitude range and are statistically significant. The dummy-variable-for-attack-helicopters coefficient estimates are not statistically significant, but we use them in estimation since they are the best linear predictors in any case. The standard errors show considerable dispersion in the data; this may be caused by differences in operating conditions, and/or by inherent differences among the helicopter models. Nonetheless, there was no pattern in the residuals sufficient for confident identification of any one particular, or any set, of helicopters as being inherently more or less expensive to maintain. We therefore apply the CERs to all the alternatives as our best estimate of the underlying relationship between O&S cost and aircraft size.

Taking together the data from the effectiveness analysis, the indirect- to direct-cost ratios, and the CERs, we arrive at an average procurement unit cost, one-year O&S cost, and present value of life-cycle costs over a 20-year lifetime for each helicopter alternative.[15]

---

[15] These are presented in Mouton et al., 2014, not available to the general public.

CHAPTER FIVE

Cost-Effectiveness Integration Analysis and Results

In Chapter Two we identified the missions that helicopter fleets of U.S. partner nations must be able to execute to achieve mutual security objectives; in Chapter Three we evaluated each platform's effectiveness in executing those missions; and in Chapter Four we estimated their total 20-year life-cycle cost. We turn now to integrating these component parts into a single, meaningful measure of cost-effectiveness, estimating how much each aircraft can accomplish per dollar spent to acquire and maintain it. This measure enables us to identify, in concrete terms, the performance and price trade-offs implicit in selecting one helicopter over another.

Cost-Effectiveness Analysis

The combination of the results of the effectiveness analysis described in Chapter Three and the cost estimates developed in Chapter Four produces approximately 22,000 cost-effectiveness scores. We reduce this number to a manageable set by dividing the helicopter alternatives into three groups, by performance. The first group contains all aircraft that proved capable of performing missions on all routes in any of the 33 partner nations in the analytical country set. The second group contains those aircraft that proved capable of completing the same routes as the Mi-17v5. The third group includes all aircraft that meet neither of these criteria; because their introduction would result in degradation of route access as compared to the Mi-17v5, none of these aircraft is considered a viable alternative for the U.S. security cooperation partner nations considered here.

This section will explain our specific implementation of this constraint. For each country of focus, we define a *reference mission* as shown in Table 5.1.

Utility Helicopter Alternative Cost-Effectiveness—Afghanistan

Recall from Chapter Two that we developed three sets of mission-tasks representative of operations in Afghanistan: those for MoD, MoI, and DVs. Cost-effectiveness analysis for the helicopter alternatives was conducted on each of these three mission-task sets, and will be presented in turn.

Afghanistan MoD

For the Afghanistan MoD missions, we worked with 18 combinations: three sets of missions, two mission types, under ISA, Hot, and Extra-Hot conditions. The two basic mission types were company deployment in a single lift, and battalion deployment in a period of daylight. The mission sets were defined such that the first contains all routes for which the Mi-17v5 is capable, numbering 17; the second contains the four of these 17 routes that require the largest

Table 5.1
Reference Missions for Analysis

Country	Reference Mission
Afghanistan – MoD	Deploy a company in a single lift, Hot conditions
Afghanistan – MoI	Resupply a police outpost in a single lift, Hot conditions
Afghanistan – DV	Move a DV contingent, Hot conditions
Iraq	Deploy a company in a single lift, Hot conditions

number of Mi-17v5s; and the third contains the ten that require the smallest number of Mi-17v5s. We refer to these sets as "All but 3," "Tough 4," and "Easy Half," respectively.

The results of the analysis are presented in Table 5.2. The helicopter alternatives are displayed on the left, followed by route deficiency; this is color-coded to indicate whether the helicopter is in the group that can complete all routes (blue), the group that can complete only those 17 that the Mi-17v5 can complete (green), or the group that meets neither of these criteria (gray). Where the helicopter falls into the gray category, the number of the Mi-17v5 routes it is unable to fly is indicated as well.

For the alternatives that can complete all Mi-17v5 routes, the next column displays the effectiveness results averaged across all 18 mission-tasks and averaged over temperature conditions.[1] All scores represent the ratio of the number of the Mi-17v5s to the number of alternative helicopters needed to complete the set of mission-tasks. Thus a score of 1.0 means the alternative helicopter has the same effectiveness as the Mi-17v5. We divide scores into three ranges, again color-coded: blue indicates that the alternative score is greater than 1.15, which means it exceeds by 15 percent or more the effectiveness of the Mi-17v5 (it can achieve with ten or fewer aircraft what the Mi-17v5 can with 11.5); green indicates a score that falls between 1.0 and 1.15; orange between 0.85 and 1.0; and gray for scores less than 0.85.[2]

The last column presents the cost-effectiveness results, using the same ratio measure and color-coding. Because the cost figures associated with a number of the alternative platforms considered here are estimates, as described fully in Chapter Four, our measure of cost-effectiveness must account for some degree of uncertainty. As such, we categorize these platforms' performances by range; alternatives that have cost-effectiveness scores that fall within 15 percent of that of an Mi-17v5 (i.e., between 0.85 and 1.15) are considered "comparably cost-effective," while those that fall above this uncertainty band have a superior level of cost-effectiveness. Platforms that fall into either of these ranges are feasible candidates for replacing the Mi-17v5.

---

[1] For a helicopter to be effective in a country, it must be able to operate year-round in a wide variety of temperature conditions. This means it should be effective on an average-temperature day, on a hot day, and on an extremely hot day. Since it is difficult to apply specific weights to each of these criteria, a simple average was used. For example, an extremely hot day may only occur 1 percent of the year, but giving it a 1 percent weight may not be accurate since there may be great value in being able to operate on any day and at any time. Specifically, having this ability would deny the enemy known windows of immunity from rotary-wing assets. Based on this, in the authors' judgment, neither a 1-percent weight nor a 100-percent weight for Extra-hot is appropriate, and therefore, for simplicity, we used weights of 33 percent for each temperature condition.

[2] The precise numerical values associated with these results are presented in Mouton et al., 2014, not available to the general public.

Table 5.2
Utility Helicopter Effectiveness and Cost-Effectiveness Scores—Afghanistan MoD

		Route Deficiency	Effectiveness	Cost-Effectiveness
			Overall	Overall
Alternative	Bell 407	14		
	Bell 429	9		
	Huey II	5		
	UH-1N	6		
	Bell 412EP	8		
	UH-1Y	5		
	CH-46E	7		
	CH-47D			
	CH-47 Intl			
	S-61T			
	S-61T+			
	UH-60L			
	UH-60M	1		
	S-92A	5		
	AW109			
	AW139			
	AW149			
	EH101	7		
	EC-145			
	AS-332L1	1		
	NH-90	1		
	EC-725	1		

NOTE: See p. 44 for an explanation of color coding.

The CH-47D performs particularly well compared to the Mi-17v5, and both it and the CH-47 International are capable on all routes analyzed. The S-61T+ and the AW139 also are capable on the full roster of routes, while the S-61T, UH-60L, AW109, and EC-145 are capable on the 17 Mi-17v5 routes. Three alternatives—the AW139, AW109, and EC-145—are of comparable overall cost-effectiveness with the Mi-17v5, with scores below 1.0, but within the uncertainty range.[3] Note that of the four helicopters that can complete all of the routes, only the AW139 does not have superior cost-effectiveness. This invites a trade-off between the lower cost-effectiveness and the greater route access of the AW139 compared to the Mi-17v5.

As a robustness check, we examine whether excluding alternatives that cannot perform on the full number of Mi-17v5 routes excludes potentially attractive candidates. If it is the case,

[3] Note that the AW109 and EC-145—as well as the other two smallest helicopters examined, the Bell 407 and Bell 429—do not carry two loadmaster/gunner personnel, or door-gun and ammunition. Thus their cost-effectiveness scores must be interpreted with this in mind.

for example, that an alternative performs exceptionally well in almost all routes but has poor performance on one of the routes that the Mi-17v5 can fly, then this aircraft might still be a viable replacement. We find this not to be the case for the helicopter alternatives studied here; in all cases, their cost-effectiveness falls below the 0.85 threshold—including for four aircraft excluded on the basis of failing to complete only one of the 17 Mi-17v5 routes. This check was run for all of the cost-effectiveness analyses that follow, and the results hold.

We next examine the data to determine whether a mixed fleet—i.e., one composed of more than one helicopter type—would be appropriate for replacing Mi-17v5s for the Afghanistan MoD. One way this might be so is if helicopter performance were differential, with two different alternatives demonstrably more cost-effective for two different sets of missions, but even this would be important only in instances in which *both* sets of missions need to be accomplished at the same time. The results do not indicate any differential pattern of performance across the alternatives, suggesting a mixed fleet would not be a necessary or particularly useful approach to achieving cost-effectiveness for the Afghanistan MoD.

We provide a second set of results to further illustrate the effects of mission characteristics on cost-effectiveness. This time, we take as our representative mission a small-payload, single daily-lift sustainment operation. The results, displayed in Table 5.3, indicate that the CH-47 is not cost-effective here. Rather, the cost-competitive alternatives are the S-61T, S-61T+, UH-60L, UH-60M, AW109, and EC-145. However, although the CH-47, AW139, and AW149 are not cost-effective, they do offer access to more locations than the Mi-17v5.

Afghanistan DV Mission

Table 5.4 shows results for the Afghanistan DV mission. The CH-47s and the S-61T+ are still in the cost-competitive set. Joining them are the EH101, AS-332L1, and EC-725.

Afghanistan MoI Missions

The results for three Afghanistan MoI missions—deploying a police unit, sustaining an outpost in a single lift, and sustaining an outpost during a period of daylight—appear in Tables 5.5–5.7. The cost-effective alternatives are broadly comparable to those for the MoD unit-deployment missions. However, the large CH-47s are not as cost-effective in the outpost daylight-sustainment mission because the movement requirements are relatively small, rendering the particular advantages of the larger CH-47 less important. The S-61T, UH-60L, AW149, and AS332L1 are all added to the cost-competitive set in the police and single-lift MoI missions.

Utility Helicopter Alternative Cost-Effectiveness—Iraq

The results for the Iraq MoD unit deployment missions are somewhat unique, in that no alternative platform achieves a superior cost-effectiveness score. But the CH-47s, the S-61Ts, and the AS-332L1 are cost-competitive with the Mi-17v5, which is capable on 17 routes here. Complete results appear in Table 5.8.

Utility Helicopter Alternative Cost-Effectiveness—Other Partner Nations

This section presents detailed results for the high-long and low-short archetype countries defined in Chapter Two for two missions: company movement and group movement.

Table 5.3
Utility Helicopter Effectiveness and Cost-Effectiveness—Afghanistan MoD, Sustainment Mission

		Route Deficiency	Effectiveness Overall	Cost-Effectiveness Overall
Alternative	Bell 407	5		
	Bell 429	1		
	Huey II	3		
	UH-1N	4		
	Bell 412EP	4		
	UH-1Y	4		
	CH-46E	4		
	CH-47D			
	CH-47 Intl			
	S-61T			
	S-61T+			
	UH-60L			
	UH-60M			
	S-92A	3		
	AW109			
	AW139			
	AW149			
	EH101	6		
	EC-145			
	AS-332L1	1		
	NH-90			
	EC-725			

NOTE: See p. 44 for an explanation of color coding.

For the high-long company-movement requirement, the Mi-17v5 can achieve completion on 18 routes. As shown in Table 5.9, five helicopter alternatives can complete these same routes, and with high cost-effectiveness relative to the Mi-17v5.

In the high-long group-movement mission set, shown in Table 5.10, the Mi-17v5 can deploy a group in a single lift under Hot conditions on 19 routes. Seven of the alternatives can do the same, six of which with superior cost-effectiveness, and one with comparable cost-effectiveness.

All alternatives except the Bell 407 can complete the same low-short company-movement missions as the Mi-17v5, but only a subset can do so with comparable or superior cost-effectiveness. The CH-47D and S-61T+ have very good cost-effectiveness, and the CH-47 International, S-61T, AS-332L1, and EC-725 are cost-competitive (see Table 5.11).

All alternatives can complete the same low-short group-movement missions as the Mi-17v5. Here, even more alternatives are cost-competitive with the Mi-17v5 (see Table 5.12).

Table 5.4
Utility Helicopter Effectiveness and Cost-Effectiveness—Afghanistan DV

		Route Deficiency	Effectiveness Overall	Cost-Effectiveness Overall
Alternative	Bell 407	8		
	Bell 429	8		
	Huey II	4		
	UH-1N	8		
	Bell 412EP	7		
	UH-1Y	5		
	CH-46E	6		
	CH-47D			
	CH-47 Intl			
	S-61T			
	S-61T+			
	UH-60L	1		
	UH-60M	2		
	S-92A	2		
	AW109	6		
	AW139			
	AW149			
	EH101			
	EC-145	1		
	AS-332L1			
	NH-90			
	EC-725			

NOTE: See p. 44 for an explanation of color coding.

Attack Helicopter Alternative Effectiveness

Table 5.13 shows the effectiveness level for the attack helicopter alternatives evaluated on a representative MoD unit-movement for each actual or archetype country. The first column of numbers indicates the number of routes in each country, the second displays the percentage of this number that the Mi-17v5 can complete. The next four columns contain the percentage of the routes in each country completed, with fire support, by each attack helicopter alternative, respectively.

Note that the AW129 and the AH-1Z have very good route access, the AH-64D has generally inferior access, and the Mi-35 has very poor access. As stated earlier, the AW129 is cost-competitive with the AH-1Z at a flyaway cost in the $40 million range while the AH-64D is cost-competitive with the AH-1Z in those cases in which it can access the routes.

Table 5.5
Utility Helicopter Effectiveness and Cost-Effectiveness—Afghanistan MoI Police Missions

		Route Deficiency	Effectiveness	Cost-Effectiveness
			Overall	Overall
Alternative	Bell 407	8		
	Bell 429	2		
	Huey II			
	UH-1N	2		
	Bell 412EP	3		
	UH-1Y			
	CH-46E	3		
	CH-47D			
	CH-47 Intl			
	S-61T			
	S-61T+			
	UH-60L			
	UH-60M			
	S-92A	3		
	AW109			
	AW139			
	AW149			
	EH101	5		
	EC-145			
	AS-332L1			
	NH-90			
	EC-725			

NOTE: See p. 44 for an explanation of color coding.

Table 5.6
Utility Helicopter Effectiveness and Cost-Effectiveness—Afghanistan MoI Outpost Missions (Single Lift)

		Route Deficiency	Effectiveness Overall	Cost-Effectiveness Overall
Alternative	Bell 407	6		
	Bell 429	3		
	Huey II	1		
	UH-1N	4		
	Bell 412EP	3		
	UH-1Y	2		
	CH-46E	3		
	CH-47D			
	CH-47 Intl			
	S-61T			
	S-61T+			
	UH-60L			
	UH-60M	1		
	S-92A	2		
	AW109	1		
	AW139			
	AW149			
	EH101	2		
	EC-145			
	AS-332L1			
	NH-90			
	EC-725			

NOTE: See p. 44 for an explanation of color coding.

Table 5.7
Utility Helicopter Effectiveness and Cost-Effectiveness— Afghanistan MoI Outpost Missions (Period of Daylight)

		Route Deficiency	Effectiveness	Cost-Effectiveness
			Overall	Overall
Alternative	Bell 407	6		
	Bell 429	3		
	Huey II	1		
	UH-1N	4		
	Bell 412EP	3		
	UH-1Y	2		
	CH-46E	3		
	CH-47D			
	CH-47 Intl			
	S-61T			
	S-61T+			
	UH-60L			
	UH-60M	1		
	S-92A	2		
	AW109	1		
	AW139			
	AW149			
	EH101	2		
	EC-145			
	AS-332L1			
	NH-90			
	EC-725			

NOTE: See p. 44 for an explanation of color coding.

Table 5.8
Utility Helicopter Effectiveness and Cost-Effectiveness—Iraq MoD

		Route Deficiency	Effectiveness	Cost-Effectiveness
			Overall	Overall
Alternative	Bell 407	6		
	Bell 429	7		
	Huey II	3		
	UH-1N	5		
	Bell 412EP			
	UH-1Y	2		
	CH-46E			
	CH-47D			
	CH-47 Intl			
	S-61T			
	S-61T+			
	UH-60L			
	UH-60M	1		
	S-92A			
	AW109	1		
	AW139			
	AW149			
	EH101	1		
	EC-145			
	AS-332L1			
	NH-90			
	EC-725			

NOTE: See p. 44 for an explanation of color coding.

Table 5.9
Utility Helicopter Cost-Effectiveness—High-Long Company Movement

		Route Deficiency	Effectiveness	Cost-Effectiveness
			Overall	Overall
Alternative	Bell 407	18		
	Bell 429	17		
	Huey II	13		
	UH-1N	17		
	Bell 412EP	18		
	UH-1Y	17		
	CH-46E	15		
	CH-47D			
	CH-47 Intl			
	S-61T	7		
	S-61T+			
	UH-60L	6		
	UH-60M	9		
	S-92A	17		
	AW109	6		
	AW139			
	AW149			
	EH101	17		
	EC-145	9		
	AS-332L1	5		
	NH-90	1		
	EC-725	3		

NOTE: See p. 44 for an explanation of color coding.

Table 5.10
Utility Helicopter Cost-Effectiveness—High-Long Group Movement

		Route Deficiency	Effectiveness	Cost-Effectiveness
			Overall	Overall
Alternative	Bell 407	18		
	Bell 429	11		
	Huey II	9		
	UH-1N	14		
	Bell 412EP	18		
	UH-1Y	14		
	CH-46E	16		
	CH-47D			
	CH-47 Intl			
	S-61T	5		
	S-61T+			
	UH-60L	7		
	UH-60M	10		
	S-92A	18		
	AW109	7		
	AW139			
	AW149			
	EH101	16		
	EC-145			
	AS-332L1	4		
	NH-90			
	EC-725	2		

NOTE: See p. 44 for an explanation of color coding.

Table 5.11
Utility Helicopter Cost-Effectiveness—Low-Short Company Movement

		Route Deficiency	Effectiveness	Cost-Effectiveness
			Overall	Overall
Alternative	Bell 407	3		
	Bell 429			
	Huey II			
	UH-1N			
	Bell 412EP			
	UH-1Y			
	CH-46E			
	CH-47D			
	CH-47 Intl			
	S-61T			
	S-61T+			
	UH-60L			
	UH-60M			
	S-92A			
	AW109			
	AW139			
	AW149			
	EH101			
	EC-145			
	AS-332L1			
	NH-90			
	EC-725			

NOTE: See p. 44 for an explanation of color coding.

Table 5.12
Utility Helicopter Cost-Effectiveness—Low-Short Group Movement

		Route Deficiency	Effectiveness Overall	Cost-Effectiveness Overall
Alternative	Bell 407			
	Bell 429			
	Huey II			
	UH-1N			
	Bell 412EP			
	UH-1Y			
	CH-46E			
	CH-47D			
	CH-47 Intl			
	S-61T			
	S-61T+			
	UH-60L			
	UH-60M			
	S-92A			
	AW109			
	AW139			
	AW149			
	EH101			
	EC-145			
	AS-332L1			
	NH-90			
	EC-725			

NOTE: See p. 44 for an explanation of color coding.

Table 5.13
Attack Helicopter Route Execution

Country	Number of Routes	Mi-17v5	AW129	AH-1Z	AH-64D	Mi-35
Afghanistan	20	85%	85%	100%	60%	40%
Iraq	17	100%	100%	100%	94%	76%
High-Long	25	76%	36%	100%	20%	0%
Low-Short	10	100%	100%	100%	100%	100%

NOTE: See p. 44 for an explanation of color coding.

CHAPTER SIX

Summary Findings

This chapter presents our summary findings. We offer our observations on the Mi-17v5 helicopter, then discuss the findings for our countries of interest and for those of the rest of the world. These results focus on the utility helicopters. Finally, we provide our findings on analysis of the attack helicopters considered.

Key Findings: Utility Helicopters

We begin by making several observations about the performance and cost-effectiveness of the reference helicopter, the Mi-17v5. The Mi-17v5 was unable to complete all of the mission-tasks on all the routes evaluated. In some cases, limitations were imposed by range; in others, by required altitude or hover capability. We note also that the Mi-17v5 generally declined in cost-effectiveness relative to the alternatives as climate conditions increased in temperature.

Our analysis indicates that among utility platforms, the Boeing CH-47D, Sikorksy S-61T, Eurocopter AS-332L1 Super Puma, and the AgustaWestland AW139 are consistently more cost-effective than the Mi-17v5. The Sikorsky S-61T+ performs similarly well, but is currently in development. In these cases, the margin of increase in cost-effectiveness over the Mi-17v5 often is such that the other aircraft achieve greater efficiency even when additional tail requirements are applied.

Several small utility helicopters also had good cost-effectiveness, including the Eurocopter EC-145 (LUH-72A Lakota) and the AgustaWestland AW109. Note, however, that in this analysis, these aircraft did not carry two loadmaster/gunner personnel, or a door-gun and ammunition. Thus, these platforms lack the defensive capability needed to suppress enemy action in the landing zone. The cost-effectiveness results presented here do not penalize for this fact, so the findings must be interpreted keeping it in mind. For attack aircraft, the AH-1Z and the AW129 are able to accomplish many of the same missions at comparable cost to the Mi-17v5. That makes these aircraft feasible candidates for providing armed escort, as well as over-watch for cargo helicopters.

These general findings are accompanied by more detailed and country-specific results of interest. We provide here a subset of findings to illustrate the significance of mission specification—changes in mission requirements can significantly affect a helicopter's cost-effectiveness.

Table 6.1 provides cost-effectiveness levels for utility helicopters completing infiltration and sustainment/CASEVAC mission-tasks representative of those undertaken by the MoDs in

Afghanistan and Iraq.[1] Each of the missions reported in Table 6.1 requires deploying military units of at least company-size from departure bases to landing zones at various locations in the country.

Because the cost figures associated with a number of the alternative platforms considered here are estimates, as described fully in Chapter Four, our measure of cost-effectiveness must account for some degree of uncertainty. As such, we categorize the cost-effectiveness of those that fall within 15 percent of that of an Mi-17v5 (i.e., between 0.85 and 1.15) as being "comparably cost-effective," and those that fall above this uncertainty band as having a superior level of cost-effectiveness. Platforms that fall into either of these categories are feasible candidates for replacing the Mi-17v5.

We divide scores into four bands—designated "superior," "strongly competitive," "competitive," and "not competitive"—as indicated by color, respectively: blue indicates that the alternative score is greater than 1.15, which means it exceeds the effectiveness of the Mi-17v5 by 15 percent or more (it can achieve with ten or fewer aircraft what the Mi-17v5 can with 11.5); green indicates a score that falls between 1.0 and 1.15; orange between 0.85 and 1.0; and gray indicates scores less than 0.85. The table makes clear that for these four key partner nations, the CH-47D, CH-47 International, and the S-61T+ are far more cost-effective than the Mi-17v5, and the S-61T and AS-332L1 are strongly competitive.

Table 6.2 presents results for the MoD unit-deployment mission alongside an MoD sustainment mission, a DV movement mission, and three missions representative of the type conducted by the MoI: deployment of a police unit, sustaining an outpost in a single daily lift, and sustaining an outpost over a period of daylight.

Because the movement requirements of the company sustainment mission (Sustain Military Units) and the outpost daylight-sustainment mission (Sustain an Outpost—All-Day Delivery) are both relatively small, the large cargo capacity of the CH-47s is not as great an advantage. Thus, it is not surprising that the smaller S-61T, which is better suited to the movement size and costs less than the CH-47s, has better cost-effectiveness in these small missions.

Similar results are produced through analysis of the cost-effectiveness for MoD company-sized unit deployment missions in four archetype countries, as displayed in Table 6.3. Here again, the CH-47D has superior cost-effectiveness, and the S-61T+ is at least competitive, across all archetypes. Notably, the S-61T+ is least cost-competitive in low-altitude, long-distance missions; by comparison, the AS-332L1 (Super Puma) is strongly competitive in these low-altitude categories.

The cost-effectiveness results for group-movement missions in four archetype partner nations appear in Table 6.4. Here, the alternatives that were competitive for the key partner nations are similarly competitive at high altitudes. In the low-altitude categories, however, results differ substantially. For these archetype countries, it is the smaller helicopters that emerge as being cost-competitive, with the S-61T+ the only platform with superior cost-effectiveness across all altitude-distance combinations. Perhaps not surprisingly, shorter missions involving small force movements allow a wider range of potentially good alternatives.

[1] Further information on the other key partner nations is provided in Mouton et al., 2014, not available to the general public.

Table 6.1
Summary of Utility Helicopter Cost-Effectiveness—MoD Mission-Tasks in Afghanistan and Iraq

		Country	
		Afghanistan	Iraq
Alternative	Bell 407		
	Bell 429		
	Huey II		
	UH-1N		
	Bell 412EP		
	UH-1Y		
	CH-46E		
	CH-47D		
	CH-47 Intl		
	S-61T		
	S-61T+		
	UH-60L		
	UH-60M		
	S-92A		
	AW109		
	AW139		
	AW149		
	EH101		
	EC-145		
	AS-332L1		
	NH-90		
	EC-725		

	Superior cost-effectiveness (CE) (CE > 1.15)
	Strongly cost competitive (1.0 < CE < 1.15)
	Cost competitive (0.85 < CE < 1.00)
	Not cost-effective (CE < 0.85)

Table 6.2
Utility Helicopter Cost-Effectiveness—Afghanistan MoD, MoI, and DV

		MoD		MoI			
		Deploy Military Units	Sustain Military Units	Deploy a Police Unit	Sustain an Outpost—Single Delivery	Sustain an Outpost—All-Day Delivery	DV Movement
Alternative	Bell 407						
	Bell 429						
	Huey II						
	UH-1N						
	Bell 412EP						
	UH-1Y						
	CH-46E						
	CH-47D						
	CH-47 Intl						
	S-61T						
	S-61T+						
	UH-60L						
	UH-60M						
	S-92A						
	AW109						
	AW139						
	AW149						
	EH101						
	EC-145						
	AS-332L1						
	NH-90						
	EC-725						

	Superior cost-effectiveness (CE > 1.15)
	Strongly cost competitive (1.0 < CE < 1.15)
	Cost competitive (0.85 < CE < 1.00)
	Not cost-effective (CE < 0.85)

Table 6.3
Utility Helicopter Cost-Effectiveness—Partner Nation Archetypes with Company Movement Requirements

		Country Type	
		High-Long	Low-Short
Alternative	Bell 407		
	Bell 429		
	Huey II		
	UH-1N		
	Bell 412EP		
	UH-1Y		
	CH-46E		
	CH-47D		
	CH-47 Intl		
	S-61T		
	S-61T+		
	UH-60L		
	UH-60M		
	S-92A		
	AW109		
	AW139		
	AW149		
	EH101		
	EC-145		
	AS-332L1		
	NH-90		
	EC-725		

	Superior cost-effectiveness (CE > 1.15)
	Strongly cost competitive (1.0 < CE < 1.15)
	Cost competitive (0.85 < CE < 1.00)
	Not cost-effective (CE < 0.85)

Table 6.4
Utility Helicopter Cost-Effectiveness—Partner Nation Archetypes with Group Movement Requirements

		Country Type	
		High-Long	Low-Short
Alternative	Bell 407		
	Bell 429		
	Huey II		
	UH-1N		
	Bell 412EP		
	UH-1Y		
	CH-46E		
	CH-47D		
	CH-47 Intl		
	S-61T		
	S-61T+		
	UH-60L		
	UH-60M		
	S-92A		
	AW109		
	AW139		
	AW149		
	EH101		
	EC-145		
	AS-332L1		
	NH-90		
	EC-725		

	Superior cost-effectiveness (CE > 1.15)
	Strongly cost competitive (1.0 < CE < 1.15)
	Cost competitive (0.85 < CE < 1.00)
	Not cost-effective (CE < 0.85)

Key Findings: Attack Helicopters

Finally, we assessed the cost-effectiveness of four attack helicopter alternatives. Recall from Chapter Three that these were evaluated for effectiveness across the same missions as the utility helicopters for all four key partner and archetype nations, and that the relevant metric was their ability to service the same routes as the Mi-17v5. When integrated with cost estimates, our model identifies the AH-1Z and AW129 as being able to access the same routes as the Mi-17v5. Recall also that we were unable to acquire reliable information on the procurement cost of the AW129, and so took as our estimate a cost of approximately $40 million; if this estimate is accurate, the AW129 is cost-competitive with the AH-1Z. The AH-64D is not well suited to the kind of long-distance, high-altitude, hot-weather missions that predominate in our four key partner nations. The Mi-35 is particularly ill-suited for these countries, and can provide fire support only for a small number of the 100 routes analyzed. These results are provided in Table 6.5.

Table 6.5
Attack Helicopter Route Execution

Country	Number of Routes	Mi-17v5	AW129	AH-1Z	AH-64D	Mi-35
Afghanistan	20	85%	85%	100%	60%	40%
Iraq	17	100%	100%	100%	94%	76%
High-Long	25	76%	36%	100%	20%	0%
Low-Short	10	100%	100%	100%	100%	100%

NOTE: See p. 44 for an explanation of color coding.

References

Aircraftcompare.com, *Agusta-Westland—United Kingdom, Military AW109 LUH,* web page, undated-a. As of December 12, 2013:
http://www.aircraftcompare.com/helicopter-airplane/Agusta-Westland-AW109-LUH/253

———, *AW139,* web page, undated-b. As of December 12, 2013:
http://www.aircraftcompare.com/helicopter-airplane/Agusta-Westland-AW139/259

———, *AW149,* web page, undated-c, UNCLASSIFIED. As of December 12, 2013:
http://www.aircraftcompare.com/helicopter-airplane/Agusta-Westland-AW149/254

———, *Military AW101 Merlin,* web page, undated-d. As of December 12, 2013:
http://www.aircraftcompare.com/helicopter-airplane/Agusta-Westland-AW101-Merlin/251

———, *Military AW129 Mangusta,* web page, undated-e. As of December 12, 2013:
http://www.aircraftcompare.com/helicopter-airplane/Agusta-Westland-AW129-Mangusta/252

ARINC, Mi-17v5 contract, W9113M–07–D–0009, October 23, 2008.

Boeing, CH-47 maintainer course program of instruction, undated.

de Briganti, Giovanni, "U.S. State Department Order for 110 S-61Ts Is Boon for Sikorsky," *Rotor & Wing,* May 1, 2010. As of December 9, 2013:
http://www.aviationtoday.com/rw/military/attack/rotorcraftreport-u-s-state-dept-order-for-110-s-61ts-is-boonfor-sikorsky_67821.html#.UpydKI7mw70

Department of the Army, Procurement Programs, *Fiscal Year 2010 Budget Estimate: Aircraft Procurement,* 2009.

———, *Fiscal Year 2011 Budget Estimate: Aircraft Procurement,* 2010.

———, *Fiscal Year 2012 Budget Estimate: Aircraft Procurement,* 2011.

DoD—*See* U.S. Department of Defense.

DoS—*See* U.S. Department of State.

FedBizOpps.gov, KC-X Tanker Modernization Program Request for Proposals (Solicitation *FA8625-10-R-6600-SpecialNotice*), web page, September 25, 2009. As of January 21, 2014:
https://www.fbo.gov/?s=opportunity&mode=form&tab=core&id=713bc6e87f1a76db2c2b20a4bee1e8a5&_cview=0

Federal Aviation Administration, Code of Federal Regulations, §91.151.

Flight International, "Brazilian Air Force Fields First Mi-35M Attack Helicopters," Flightglobal website, April 22, 2010. As of December 12, 2013:
http://www.flightglobal.com/articles/2010/04/22/340973/pictures-brazilian-air-force-fields-first-mi-35mattack.html

Grissom, Adam, Alexander C. Hou, Brian Shannon, and Shivan Sarin, *An Estimate of Global Demand for Rotary-Wing Security Force Assistance,* Santa Monica, Calif.: RAND Corporation, 2010. Not available to the general public.

Iowa Environmental Mesonet, ASOS/AWOS data download, undated. As of October 10, 2012: http://mesonet.agron.iastate.edu/request/download.phtml?network=AF_ASOS

Jane's All the World Aircraft, *Eurocopter EC 225 and EC 725,* web page, November 20, 2009. Website is subscription-only.

Mouton, Christopher A., David T. Orletsky, Michael Kennedy, Fred Timson, Adam Grissom, and Akilah Wallace, *Cost-Effective Alternatives to the Mi-17 for Partner Nations: Focus on Afghanistan, Iraq, Pakistan, and Yemen,* Santa Monica, Calif.: RAND Corporation, 2014. Not available to the general public.

NSRW Working Group deliberations, SOCOM Global Synchronization Conference, MacDill Air Force Base, Fla., 2010.

Office of Management and Budget, *Discount Rates for Cost-Effectiveness, Lease Purchase, and Related Analyses,* OMB Circular No. A-94, Appendix C, December 2011. As of October 10, 2012: http://www.whitehouse.gov/omb/circulars_a094_a94_appx-c/

Office of the Secretary of Defense, Cost Analysis Improvement Group, *Operating and Support Cost-Estimating Guide,* May 1, 1992.

Shelby, Richard, U.S. Senator, letter to Secretary of Defense, October 21, 2009.

U.S. Air Force, *AFI 65-503, Cost Factors,* web page, undated. Not available to the general public.

———, *Mi-17 Program of Instruction,* USAF Air Education and Training Command, 2012.

U.S. Army Operating and Support Management Information System (OSMIS) website, undated. Not available to the general public.

U.S. Department of Defense, *Medium-Lift Helicopter Requirements for the Afghanistan National Army Air Corps (ANAAC),* Operational Requirements and Analysis Document, February 11, 2010a, p. 13.

———, *Mi-17 Helicopters,* report to Congress, March 23, 2010b.

U.S. Department of Defense, Defense Acquisition Management Information Retrieval (DAMIR), *Selected Acquisition Report, RCS: DD-A&T(Q&A) 823-182 (LUH),* December 31, 2009a.

———, *Selected Acquisition Report (SAR): RCS: DD-A&T(Q&A) 823-278 (CH-47F),* December 31, 2009b.

———, *Selected Acquisition Report (SAR): RCS: DD-A&T(Q&A) 823-341 (UH-60M),* December 31, 2009c.

———, *Selected Acquisition Report (SAR): RCS: DD-A&T(Q&A) 823-831 (Longbow Apache),* December 31, 2009d.

U.S. Department of Defense Standard Practice, *Glossary of Definitions, Ground Rules, and Mission Profiles to Define Air Vehicle Performance Capability,* MIL-STD-3013, February 14, 2003.

U.S. Department of State, *Aviation Services Study,* March 9, 2009.

U.S. Department of State and Department of Defense, *Foreign Military Training and DoD Engagement Activities of Interest 2011–2012,* Joint Report to Congress, 2012.

U.S. Navy Visibility and Management of Operation and Support Cost (VAMOSC) website, undated. Not available to the general public.